Buzz

Urban Beekeeping and the Power of the Bee

Lisa Jean Moore and Mary Kosut

NEW YORK UNIVERSITY PRESS
New York and London

NEW YORK UNIVERSITY PRESS
New York and London
www.nyupress.org

References to Internet websites (URLs) were accurate at the time of writing. Neither the author nor New York University Press is responsible for URLs that may have expired or changed since the manuscript was prepared.

Library of Congress Cataloging-in-Publication Data
Moore, Lisa Jean, 1967–
Buzz : urban beekeeping and the power of the bee / Lisa Jean Moore and Mary Kosut.
pages cm   Includes bibliographical references and index.
ISBN 978-0-8147-6306-3 (cloth : acid-free paper)
ISBN 978-1-4798-2738-1 (paper : acid-free paper)
1. Urban bee culture—New York (State)—New York. 2. Beekeepers—New York (State)—New York—Biography. 3. Honeybee—New York (State)—New York. 4. Bee products—New York (State)—New York. 5. Bee culture—United States. 6. Honeybee—United States. 7. Honeybee—Social aspects—United States. 8. Honeybee—Effect of human beings on—United States. 9. Human-animal relationships—United States. I. Kosut, Mary. II. Title.
SF524.52.N7M66 2013
638'.1092097471—dc23 2013015223

New York University Press books are printed on acid-free paper, and their binding materials are chosen for strength and durability. We strive to use environmentally responsible suppliers and materials to the greatest extent possible in publishing our books.

Manufactured in the United States of America

c 10 9 8 7 6 5 4 3 2 1
p 10 9 8 7 6 5 4 3 2 1

Also available as an ebook

For Mike and the GCA—Mary Kosut

For Paisley, Grace, Georgia, and Greta—Lisa Jean Moore

CONTENTS

ACKNOWLEDGMENTS

As a truly collaborative project, we have worked together by immersing ourselves for more than three years in the world of beekeeping in New York City. But the collaboration goes well beyond the work of two sociologists; *Buzz* is made possible through the intra- and interspecies minglings around and beyond the metropolis. This project could not have been possible without the urban beekeepers who have been open, patient, generous, and compassionate teachers guiding us through the world of honeybees. We are especially grateful for the generosity of Liane Newton and Jim Fischer, for their taking the time to introduce us to the world of NYC beekeeping. The NYC Beekeeping Meetup is a marvelous resource for New Yorkers, providing free beekeeping instruction. We also owe much to Sam Comfort for teaching us about top bar beekeeping and providing access to his bee yards. To the honeybees, the tens of thousands that we have observed, touched, rescued, smooshed, moved, swatted, photographed, and written about, we thank you.

Our colleagues both at Purchase College and beyond have provided us with exceptional critiques and constructive advice: we are indebted to Monica Casper, Patricia Clough, Mobina Hashmi, Matthew Immergut, Chrys Ingraham, Kristen Karlberg, Suzanne Kessler, Jason Pine, and Jennifer Terry. Our colleagues Derek Denckla, Abou Farman, Shaka McGlotten, Lorraine Plourde, Michelle Stewart, and one of our students, John Comitale, thoughtfully kept all things bee on their radars by sending us bee articles, bee poetry, and bee ideas. Eben Kirksey has been invaluable as a dedicated reader and engaged scholar; this book is greatly enhanced because of his attention. Our students Rory Aledort and Lara Rodriguez have also been careful and smart readers. In particular, Mathew McDowell offered exquisite suggestions throughout the entire text and more than once corrected both animal facts (about penguins) and organizational difficulties. The Purchase College librarian,

Darcy Gervasio, provided us with etymological research assistance as well as moral support. Finally, Marcello Marcoccia's translation and astonishing interpersonal skills made access to and interviewing of Italian beekeepers possible.

Lisa Jean would like to thank her parents, Linda and Richard Moore, for their encouragement and love. Her husband, Paisley Currah, could not be a more devoted reader, enticing distraction, and doting father. Every day he reminds her of what is important in life and how to slow down to notice it. Grace, Georgia, and Greta are and will always be the most enthusiastic research assistants, intuitive sociologists, and inspiring daughters. In particular, Lisa Jean wants to apologize for not taking as seriously as she should four-year-old Grace's bee sting. (She feels, through this project and after suffering a few stings herself, overwhelming maternal guilt.) Fraser Currah provided a breadth of bee-related links and articles. She also thanks her friends Patti Curtis and Patty Howells for reading earlier chapters of the book and offering moral support.

Mary would like to thank her mother, Elizabeth Jesella, for her endless enthusiasm and morale building, as well as expert copyediting skills that tightened up this and many other projects. Her father, John Kosut, taught her early on that bees are good for us, plus he gave her vintage *National Geographic* collections and books about the end of the universe. Mary is truly grateful to her partner/pandrogyne, Mike Schreiber, who patiently engaged in long conversations about bee diseases, the merits of Joseph Beuys's artwork, and other idiosyncratic bee facts and uses. Mike's way of seeing and entangling with a slimy, creepy, weird world makes each day make more sense.

This research has been supported by the Purchase College Green Fee Fund to create a pollination garden on campus. In addition, Mary was awarded the Peter and Bette Fishbein Junior Faculty Research Award (2011–2013), which supported publication costs. We also thank Christine Dahlin for her masterful copyediting. Finally the editorial assistants Aiden Amos and Caelyn Cobb at New York University Press have provided crucial expert assistance with images, formatting, and production. Our editor, Ilene Kalish, has literally and deeply considered every word, image, argument, and concept in this book. Through her commitment to our work, this book has been enriched and transformed—we are grateful for her brainpower.

1

Catching the Buzz

Introduction

As long-term New York City residents, neither of us would consider ourselves to be huge animal lovers or nature enthusiasts. We go on occasional hikes or camping trips to escape the city, but there is always something strangely comforting about driving back into the metropolis and feeling the energy of the city—the architecture, the noises, and the people. We have both been shelter-pet owners at different stages of our lives and while trying to provide our pets with the best homes possible, there was a clear and sharp division between humans and animals: "no dogs on the bed."

As sociologists, we are invigorated when we find ourselves immersed in human subcultures where seemingly "abnormal" things happen. We like to watch members of our species pushing their bodies beyond their limits, resisting social expectations through creative problem solving—forcing us to reconsider the everyday taken-for-granted ways of life. Our previous research projects were not very engaged with animals. Mary's work interrogates artworlds, particularly how and why humans create things that come to be viewed as art and the difficulties working artists endure in the process of getting critical and economic recognition, as well as practices of body modification, from tattooing to more extreme procedures. These field sites appeal to her because they speak to the creative impulses of our species, and exploring practices and behaviors deemed to be "deviant" or "outsider" provoke us to reflect on conventional and habitual ways of being in the world. On the other hand, Lisa Jean's work focuses on human bodies in biomedical, sexual, or reproductive situations. With studies about human sperm to child sex predation to female genital anatomy and safer sex practices, her scholarship is also decidedly humancentric. Her research and projects have all examined human bodies as essential to the shape and functioning of social life. She has shown that while some bodies are highly visible others are not represented at all, with important consequences for social action and policy. In short, we know how to research humans sociologically. So for this project, we began with people.

In late 2008 through early 2009, as both of us were coming off of other scholarly projects, we began to notice the growth and intersection of do-it-yourself (DIY) urban communities, urban homesteading movements, and back-to-the-land postcollege internships. We live in Brooklyn, which is perhaps ground zero for DIY cultures; it also is a commerce-driven area that emphasizes the anti-brand or the one-of-a-kind object. Many of our former students from Purchase College seemed to follow a path of either moving to Brooklyn and getting involved in some cooperative living arrangement that involved gardening, bartering, or crafting or moving to some tristate-area farm for a lowly paid internship with organic farmers. Just as this postcollegiate shift was happening, we began to notice and hear a lot about bees. As collective artisanal beer brewing, knitting, mushroom foraging, and bread baking became points of reference while socializing, beekeeping became all the rage.

The local cultural zeitgeist seemed to pivot around quirky hands-on activities and esoteric knowledge. The act (and craft) of beekeeping if you will—nurturing teeming boxes of insects, dressing in protective gear that could be mistaken for kitsch Devo spacesuits, and harvesting "homemade" jars of honey—fits effortlessly among other eccentric eco-activities. As part of the larger 99 percent, beekeepers, rooftop farmers, and DIY crafters alike acquire a particular kind of *cultural capital*, a term the French sociologist Pierre Bourdieu used to describe networks of social relationships where cultural knowledge itself becomes a source of power and status. This type of oppositional cultural capital also offers a tangible corrective to a pop media landscape where spray-tanned, bedazzled, and shopaholic twentysomethings dominate. As the surreal and widely popular consumer orgy climaxed on the cable show *Jersey Shore*, locally owned bars and coffee shops in Clinton Hill, Brooklyn, and in the Riverside section of the Bronx hosted events organized through Facebook for knitting circles and beer brewing. Urban beekeeping overlapped with much of what we had been noticing—a shift toward bringing the "natural" into the urban, so that young people could learn a craft or a skill that seemed somehow nostalgic and lost, thereby bonding and socializing over shared interests in a seemingly disparate urban landscape and a larger cultural mediascape "gone wild."

Making our work known in our social lives, friends, or friends of friends, who had decided to keep bees started approaching us with increasing frequency. We sampled the growing array of local honey sold at the farmers' markets and compared the beeswax candles that adorned the Brooklyn Flea booth. We heard a lot of stories about bees, from news reports on Colony Collapse Disorder (CCD) to anecdotes about swarms at local spots. In short, we tingled with the buzz over bees. Like a contact high, that experience of being sober but feeling altered when surrounded by others who are chemically affected, we were catching a buzz from other people and from the bees. This buzz, a rhythmic and hypnotic sensation, an emotional and physical shift in our consciousness, resounded throughout our research journey and percusses through this book.

We wanted to learn about the beekeepers themselves, their interactions with each other, how they are part of various craft and green DIY movements, and, of course, their relationships to their bees. Who were these fellow urbanites that took on the hobby of raising bees? What motivated them? How did they become proficient? What was the phenomenology of keeping bees? We discovered that New York City beekeepers are a pleasantly motley crew of people, embodying different personal styles, political perspectives, and manners of beekeeping. The one feature that connected them was their role as the bees' stewards—they were earnest, serious, and deeply committed to their bees.

In order to learn about the practice of beekeeping and gain entry into the world of urban beekeepers, in January 2010 we enrolled in a six-month class on urban beekeeping at the Central Park Arsenal. As two novices with limited animal/insect husbandry experience, we were introduced to a part of the burgeoning New York beekeeping world and to a larger community of people who are attracted to and participate in urban homesteading and the greening of city rooftops and backyards. Jim Fischer of New York City Beekeeping led the biweekly classes, which each consisted of a two- to three-hour comprehensive PowerPoint presentation and an extensive question-and-answer period. We were quickly swept up in Fischer's sharp wit and engaged lecturing style. Always wearing a bee-themed t-shirt, Jim, a fiftyish man with gray, curly hair, became more than an instructor. Over time, he became an invaluable informant, taking us on site visits to Van Cortlandt Park or Randall's Island in his aging Volvo station wagon. Putting us in mind of a favorite uncle, he shared warmth and wisdom. He is generous with his vast knowledge and at the same time not reserved about his strong opinions of proper bee care based in science. Part of the New Alchemy Institute in the 1970s, he worked on ecological innovations in hydroponics, aquaculture, and wind power. A beekeeper for more than twenty years, before recently moving to New York City, Fischer had six hundred hives in the Blue Ridge Mountains of Virginia. Fischer combines a folksy astonishment with "big city ways" with the honed expertise of a professional and compassionate beekeeper.

Being a beekeeping student was a role we both looked forward to, as full-time professors accustomed to taking center stage in a classroom. But we were at the same time students and ethnographic researchers—the classroom as field site is a complicated social space to investigate. Our practice of simultaneously taking notes about bees combined with learning how to become a beekeeper, while also attempting a sociological meta-level analysis of who was in the room and thinking about their concerns and connections to other humans and European honeybees, was a challenge. We wanted to learn the language of bee care, understand the creatures' habits through human translations and mediations (as the bees were absent and not speaking for themselves), and at the same time get a sense of the humans who surrounded us. After six months, we had acquired some basic bee biology and beekeeping knowledge, as well as a developing sociological sketch of the human participants.

Classes began in January and usually took place on Sunday mornings that were unusually icy, windy, and frigid. Undeterred, about eighty to one hundred enthusiastic novice and seasoned beekeepers braved the weather alongside us until spring arrived, as did the bees. We were so impressed with Jim Fischer's dynamic, thorough, entertaining, and engaged lectures, that we received a grant from Purchase College, where we teach, to invite him to give a presentation entitled "Pollen Nation," attended by over 150 students and college faculty. This lecture was part of our efforts to institute a native pollinator plot at Purchase College, where we established solitary bee tubes in the spring of

2011. Our project was fueled by Purchase College's commitment to sustainability and the continuation of an on-campus garden that included common plants that attract bees. Our grant added a sustainable native bee garden plot to Purchase's campus community garden. Our plot and the placement of bees could encourage the growth and proliferation of "citizen scientists" at the undergraduate level. This plot would enable the Purchase community—faculty, staff, and students—to be part of larger National Citizen Scientists Projects as outlined by the North American Pollinator Protection Campaign or by more local organizations like the Great Pollinator Project. Citizen scientists conduct bee censuses after identifying the species of bee that is present.

However, our proposal was met with some trepidation among the members of the funding committee. Before we were to receive funding, we had to answer questions such as "How many bees are on campus?" and "What are the risks of increased bee stings due to the plot?" The fear of bees looked to be an impediment to our securing the funds. The types of questions we were asked contained, what we felt was, a degree of unanswerable panic that administrators (and others) commonly conjecture. We wrote detailed responses in hopes that we could allay fears that could prevent funding and support.[1] The difficulties we had setting up this simple garden plot really speak to how bees can instigate fear, whether they are "killer" bees or just mason or leafcutter bees. Commonly found in gardens and considered friendly or nonaggressive to humans, leafcutter bees are not described as social (as is the case with honeybees) because they make individual nest cells for their larvae, rather than within a hive. They can be introduced in gardens in cardboard tubes, what some refer to as bee hotels. The solitary bee tubes we brought into the garden are pictured in figure 1.1. The bees assimilated onto campus uneventfully and no stings were reported, administrative concerns notwithstanding. The experience of setting up these hotels gave us a productive pause, realizing just how comfortable we were feeling around bees and how familiar these insects had become to us. They were not a flying "other" who may be dodgy or even hazardous, but bees, creatures that share our space and whom we are mindful of. Although it was plainly obvious that many humans did not share our point of view—because, for them, bees are stinging insects first and foremost—we discovered bee fear as well as bee love.

Figure 1.1. Solitary bee tubes at the Purchase College Garden. (Photo credit: Mary Kosut)

In addition to our beekeeping classes and bringing bees to campus, we also attended supplemental lectures by noted bee scientists such as the entomologist Thomas Seeley, who spoke on the topic of "Honeybee Democracy: How Bees Choose a Home." We were trained as Citizen Scientists at the Brooklyn Botanic Garden for the Great Pollinator Project in the summer of 2010 and participated in an online bee census. We attended lectures on basic beekeeping at Brooklyn Brainary and Eagle Street Farms, in Greenpoint, Brooklyn. The Brooklyn Brainary is a type of urban DIY educational collective where members of the public can take classes on crafting, cooking, and special topics within the arts and sciences. The lecture we attended was on a new movement called "backwards beekeeping," a phrase coined by a charismatic L.A. beekeeper named Kirk Anderson. Both Anderson and Sam Comfort of Anarchist Apiaries in the Hudson Valley discussed noninvasive and natural approaches to beekeeping based on their experiences in urban and rural settings. Comfort, though very different from Jim Fischer, became another of our primary informants. Comfort is charismatic, like Fischer, but is more wry, even playful. He is a self-described anarchist but has a Zen quality and easy nature that suggests an accessible anarchy. Comfort, very serious about his politics, presents his radical ideas about capitalism and argues that "most bees are in a welfare state." He is in his early thirties but has a sartorial style that subverts any one particular look that may symbolically indicate a place for him within "normal" adult professional life. Comfort's clothes appear to be literally thrown on his body without looking—he is so unfashionable that he is stylish. Comfort has bee tattoos on his arms, his hair is permanently overgrown and messy, and he prefers going shoeless, advocating barefoot beekeeping.

He welcomed us into his home (a roomy two-story farmhouse filled with instruments and ephemera he shares with other young homesteaders) and his hives in upstate New York, socialized with us in Brooklyn, and explained his complete devotion to the bees in great detail. Like Fischer, Comfort has vast experience with different beekeeping practices and a fascinating biography. While he went to college for art, Comfort has a nine-year beekeeping history, starting out as a commercial beekeeper in Montana, then trucking bees to the almond fields of California and to the Pacific Northwest to pollinate cherries. After working in commercial beekeeping for four years he found that while beekeepers do all they can to keep bees alive, "monoculture is not conducive to bees' health" and "there is no young blood in the industry, no money in it, and nobody wants to get stung." Comfort has been treatment-free (not using antibiotics or pesticides) for six years and tends four hundred hives up and down the East Coast.

In the Field: Hive Checks and Human/Insect Participant Observation

In the summer of 2011, we visited Eagle Street Farms, a large rooftop vegetable farm in an industrial building in Greenpoint, Brooklyn, adjacent to the East River, on a number of occasions, to hear public lectures by Annie Novak, Meg Paska, Tim O'Neal, Sam Comfort, and other beekeepers. At public lectures, all beekeepers discussed the idiosyncratic aspects of tending bees in New York City, encouraging interested attendees to jump in and give it a try themselves. Joining Fischer and Comfort as our key informants on urban beekeeping, Meg Paska has experience beekeeping in several boroughs of the city. A redheaded twentysomething, usually dressed in a uniform of denim baggy overalls, Paska is also one of the innovators of urban homesteading—keeping chickens, gardening, and building a community of urban naturalists. Warm and funny, Paska is seriously committed to bees and has made a career as a beekeeper, for herself and others, and is also a blogger and writer working on a book about bees. Coming to beekeeping through beer-brewing workshops and an interest in creating a self-sustaining business model, Paska shared recipes with us for making mead, propolis tinctures, and beauty treatments while her rambunctious chickens trampled around the mini-farm she tended in her modest Greenpoint backyard.

We also had the pleasure of participating in hive inspections at Added Value Red Hook Community farm (also in Brooklyn), which is a large plot of land about the size of a city block that serves as a vehicle for encouraging city kids and young people to become involved in a hands-on community project. Added Value invites them to discover and appreciate the natural world in a neighborhood that is otherwise filled with housing projects and commercial development. Added Value is within a block of the new IKEA store on the waterfront, an unlikely juxtaposition of green space and landscapes of consumption.

Because we were interested in learning more about beekeeping in rural environments, in May 2011, we visited Comfort's Anarchist Apiaries, which is situated on a small collective farm and includes bee yards in other locales in Germantown, New York, about sixty miles from the city. We participated in Beekeeping Bootcamp, which included total immersion in the field (including trekking through many muddy fields that swallowed and dislodged our shoes). This weekend consisted of meeting fellow beekeepers interested in Sam's top bar hives and a non-invasive approach to beekeeping. We had our share of handling bees and managing hive expansion (including getting stung a number of times) and hearing from other beekeepers about their practices. Figure 1.2 is from this field visit. Here, Lisa Jean, with hive tool in hand, handles a top bar hive loaded with active bees and honey.

Figure 1.2. Sam Comfort's Beekeeping Bootcamp in the Upper Hudson Valley. (Photo credit: Mary Kosut)

Through the three-year period of our research we visited hives in Brooklyn (in the neighborhoods of Boerum Hill, Crown Heights, Flatbush, Bushwick, and Red Hook), in Manhattan (on Randall's Island and on the Lower East Side), and in the Bronx (at Van Cortlandt Park). We also had the opportunity to interview a beekeeping family in Southern Italy. We interviewed more than thirty urban beekeepers, from novices to seasoned beekeepers with over twenty years of experience. We interviewed hobbyists, urban homesteaders, and those who had been employed in industrial beekeeping.

Attending special events like the Inaugural New York City Honey Fest in Rockaway Beach (September 2011) and the Brooklyn Botanic Garden's "Bee-Day" Party in Prospect Park (June 2010) enabled us to witness how the culture of beekeeping has become integrated into the vibrant street-festival life of New York. At both of these events, curious members of the public attended workshops, participated in honey tastings, and observed live bees at work on their combs. We were struck by the interest and enthusiasm for bees—from how pollen can be used to cure ailments to legislation calling for New York State honey regulations—as well as the diverse ages of the attendees, including families and young adults. Such events speak to the increase in the popularity of bees, and how bees have become more visible in urban environments. Of course, bees have always thrived in New York City, but they were relatively invisible to those outside of the beekeeping community until a few years ago.

We also participated in numerous hive inspections. This allowed us not only to learn the basics of hive maintenance but also to observe the divergent ways that people "keep" bees and how they interact with them. We quickly found out that while there are common tasks to be performed, such as checking to see if bees are healthy and have enough space to live, each beekeeper has his or her own philosophical and emotional relationship with the bees. In this context, each inspection we did was colored by the beekeeper, who set the tone and the rhythm for our encounters with these unfamiliar insects. For example, some suited up in full gear and asked us to limit our conversation and quiet the tone of our voices so as not to disturb the bees (see figure 1.3). In these instances, we respected the work from a distance (usually a few feet away) and had a chance to see how beekeepers move and embody their practice. We got close to the hives themselves when we were invited to do so or when it felt unintrusive to poke our heads close to the humming bars of wax, comb, and honey. Other beekeepers, often unprotected by veils and suits, ungloved, and excited to show us their hives, commanded us to dive in (sometimes it felt a little like a dare).

After getting over our neophyte fears, we slowly adapted to handling and moving screens of bees around. We got used to being dive-bombed in the head, and we reacted calmly when a bee took a temporary seat on a wrist or pant leg. We were excited to stick our fingers in juicy combs of honey while bees were still frenetically and energetically making it.

Because we could not interview bees through traditional face-to-face conversations, our interactions with this species took the form of active observations and most specifically through these hive checks. As with our previous research projects, these subjects, bees, don't require Institutional Review Board (IRB) approval or consent forms. So we did not get IRB approval for the bees. The IRB exists to ostensibly protect human subjects from physical or emotional harm or from slander

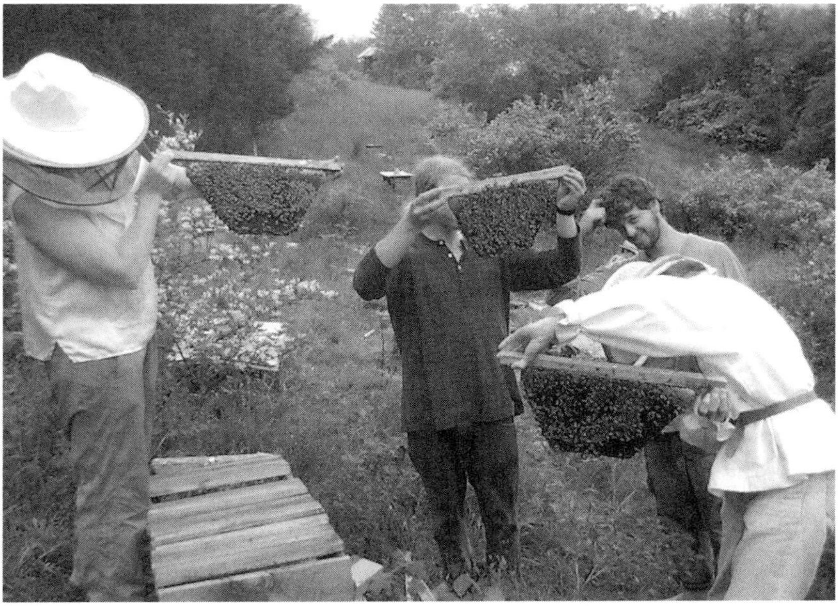

Figure 1.3. Beekeepers inspecting hives with and without gear: we have veils and jackets while others are unprotected. (Photo credit: Mary Kosut)

or defamation of character.[2] Even though human research subjects have been abused throughout human history, there have been regulations that have been refined since the Nuremberg trials.[3] This U.S.-administered military tribunal opened criminal proceedings against twenty-three leading German physicians and administrators. The Nazi physician Josef Mengele's infamous experimentation, particularly on children's bodies, is often cited at the Nuremberg Doctor's Trial of 1946–1947.[4] The creation of the Nuremberg Code of Ethics adopted in 1948 was the first modern legal attempt to establish ethical standards for modern bioscientific research, and the effects have been far-reaching. As the scope of governmental funding of bioscientific research grew, so, too, did the creation of review boards for government-sponsored research—these boards were created in hundreds of institutions. In 1979, the National Commission for the Protection of Human Subjects of Biomedical and Behavioral Research Commission published recommendations known as the Belmont Report.[5] In order to conduct research at most universities, like our own, scholars must undergo training (web-based or in the classroom) regarding the Belmont principles. Even though we as social scientists are not conducting *biomedical* research on human subjects, any interaction with human subjects generally falls under the IRB purview. Certain beings get extra consideration in social scientific research and other beings go unrecognized as being worthy of protection.

Human subjects are defined as living individuals about whom an investigator obtains (1) data through intervention or interaction with the individuals or (2) identifiable private information. All these regulations and consent forms imply that humans are vulnerable in social scientific research. But we also found that the bees are vulnerable, too: while in the process of participating directly in hive checks it is easy to crush and kill bees while moving frames, disrupting their homes. Starting with the ability to locate the queen, identify the drones, and watch new bees being born, we were soon able to assist beekeepers and the bees. We added frames and boxes to existing hives to increase the size of the bees' living environments. These hive checks were often the beekeepers' way of controlling swarms and inspecting the bees for mites, such as *Varroa destructor*. We participated in sugaring the bees, a less invasive way of managing *Varroa* and other mites by sifting powdered sugar over the bees, which encourages them to clean themselves off, hopefully disengaging any unwanted invaders from their bodies. Bees landed on our skin, stung our bodies, bounced off of our faces, and got entangled in our hair. We accidentally stepped on bees—something we never experienced before in fieldwork, effectively killing our research subjects—and delightedly rescued them from water bowls, again a new feat to save our research subjects. On occasions we would just "hang out with the bees," watching them on their landing strips taking off and returning from their foraging expeditions. It was a relaxing fieldwork experience once we got over our initial anxieties about the bees, and we learned to move in ways that lessened their anxieties about us. For example, we did not eat bananas before tending hives because a threatened bee releases an alarm pheromone that is a chemical also found in bananas. This misperceived scent could trigger bees to attack in a defensive maneuver.

There is a process called "going native" when doing qualitative fieldwork. This happens when the researcher who is participating in the field site becomes so enmeshed in the community that she can no longer maintain the personal and intellectual distance from those being studied. The concern about "going native" is that researchers will overidentify with the other human subjects and lose their critical stance. They will no longer notice how certain social practices that they studied are in any way noteworthy. Although we can very handily assist during a hive inspection or honey extraction, we want to make it clear that we have not become beekeepers. But we have developed an extremely different relationship to bees than when we first began this project more than three years ago.

There are few scholarly or scientific concerns that human researchers could go native over their engagement with another species. We could not see, act, or communicate like bees, so the risk of our adopting their worldview was impossible. However, there is a sense that our "native" humanness is decentered through our engagement with the bees. Indeed when discussing our research with others, it is always dicey when we seem to hold an empathetic stance toward the Africanized honeybee or when we consider the seemingly unavoidable bees' suffering in exchange for human survival.

Our lives, bodies, flesh, thoughts, and dreams have been transformed through *Buzz*. There are rather hypnotic hums that ricochet among fellow humans, bees, and the city and that create intense vibrations all around us. Urban dwellers can feel the subway beneath and the helicopter above: the city never slows, it buzzes with life all around us. We were attuned to these material objects grumbling in our city, but we now take notice of other inhabitants previously taken for granted. We want to share beekeepers' and our own euphoria of buzzing with feelings, sensations, and ideas through interspecies mingling. Although we have our comfort zones interviewing and observing humans, bees now demand our attention. This book attempts to describe these sonic and vibrating exchanges among organisms (humans, bees, *Varroa*), which are often obscured by ideas of being human. As this book explores, bees are integral to human life and their buzz ripples into an astonishing array of human necessities and enterprises. Our challenge has been to be as diligent as the bee, staying on task of the "main idea" of this book (for our fellow humans) while also using our conceptual toolbox to make sense of life in a multispecies world.

Because this is an api-ethnography that relies heavily on our experiences in the field, fieldnotes will be incorporated throughout the text as well as informants' own descriptions, which we have transcribed. We use fieldnotes, usually in italics, in order to represent the immediate and informal, as well as less mediated experiences we had encountering the bees. We follow the buzz, like a luminous line, throughout the pages that follow. The buzz streaks through what we have learned from the bees and their keepers, and it is the organizing principle of this book. We hope that you will also experience in a way the endorphin rush and exquisite pain of petting the bees, standing in the middle of the hive in flight, and getting stung.

2

Buzzing for Bees

From Model Insect to Urban Beekeeping

Sun has been relentless and angry all day. Everything is too hot to touch and radiating energy, the door to my apartment, the car steering wheel, the sidewalks. I feel the metal of the subway turnstile through my t-shirt, warm and greasy, probably lubricated by the sweat of a few million neighbors. Meeting some beekeepers in Brooklyn at 7, no reprieve from the heat. Everyone at the house is well covered in buttondowns and pants tucked in preparation for the hive check. They have an extra veil for me, which I toss on for the hell of it to protect my long sweaty hair from trapping bees. Everyone puts on bright green plastic gloves. All looking like characters from a 70s Woody Allen movie, grabbing the tools, metal smoker, suited up in out-of-date space-age outfits. Why do people want to do this?

We are very quiet as we climb the two or three flights of stairs to get to the roof. Faces red and panting. Door opens and feet squish, literally sink into, the rooftop. Like standing on a warm scratchy tar-paper cookie. Manhattan skyline opens up in the distance, and we survey the more mundane buildings, and the elevated LIRR train that snakes through Brooklyn. Views slow us down for a moment. It is high summer in the city. Squinting, panning around 360 degrees and there is the white box, the beehive. It is surrounded by vents and chimneys and doors, plastic ripped-up kiddie pools, dead mangled antennas, bleached-out broom handles, satellite dishes, leaves, cig butts, paint cans, and, yes, flying bees. No people around but us, standing in an archaeology of rooftop detritus. I can't help wondering, is this a good place for bees to live? Are they OK up here in this elevated wasteland? Definitely not hospitable to humans for long, but when the hives are opened we find that the boxes are full of brood, active bees, and fresh warm honey. So the bees can live on top of the city . . .

—Crown Heights, Brooklyn, June 2011 (see figure 2.1)

New Encounters with Insects

There are approximately 230 different species of bees living in the greater New York metropolitan area.[1] As the city swarms with human activity, these bees quietly pollinate fruits, vegetables, plants, and wildflowers, playing an integral part in the local urban ecology.[2] Bees have always lived throughout the five boroughs of New York with or without the aid of humans. Yet, until very recently, most people never thought of them as a species that "naturally" belonged in the city. Bees don't spring to mind when we think of urban wildlife. Pigeons, in comparison, are nonhuman fixtures within urban landscapes—they are often used to symbolize the city, making cameos in films and tourist photos. Annoying as they may be, we recognize these critters as part of the metropolis—"rats with wings," as Woody Allen quipped. Honeybees don't appear to be a natural fit for New York City. They conjure up a more bucolic environment because they are seemingly more at home within fields and farms, rather than on endless rooftops of apartment buildings.

Beekeeping has become an increasingly popular practice, and signs indicate that bees are welcomed and embraced across different neighborhoods that vary economically and culturally—they are tended atop the Whitney Museum of Art in Manhattan and in community garden plots in Bushwick, Brooklyn. Bees' new status in the city was heralded in a *New York* magazine article covering "the everything guide to urban honey," which advised readers to "think of them as your new pets."[3] Gotham is literally buzzing about bees, once a rather taken-for-granted species. In general, people have become more aware of their presence, but even more significantly, beehives are being set up in unexpected metropolitan places.

We found ourselves drawn to this buzz, slowly at first, but then it was impossible to ignore. At the end of our first introductory lecture on urban beekeeping in late January, during a chilly, speedy walk to the subway, we shared how surprised we were to see so many people "into bees." We wondered aloud, "Don't these people have a life? How can they be so wrapped up in the lives of bees?" From an outsider vantage point, it seemed like some kind of an insect cult meeting with

Figure 2.1. Inspecting frames on a Crown Heights roof. (Photo credit: Lisa Jean Moore)

attendees on their edge of their seats, taking notes, and asking such detailed questions as if the life of the species itself was in their hands. The problems of living in New York City, like escalating rent and subway fare increases, as well as larger national social and political issues like inflation, debt, and unemployment, temporarily disappeared and our concern was directed toward the bees, their needs, and our ethical responsibility to them. Was keeping healthy bees that important in the context of so many consequential and seemingly unfixable human problems? Yet, at the same time, we were engaged in the lecture and it became clear to us that we had no clue as to how complicated the bees' networks were—their relation to each other as members of a hive, to other hive colonies, to other species, to the local flora and fauna, from cherry trees and milkweed, to mites and wasps and back to humans. The relationship between beekeeper and bee was sociologically intriguing, but we were also swept up in learning more about the bees themselves, including their anatomy, method of reproduction, how they ate and excreted waste, why and how they made honey, and what they needed to live productively in the Manhattan area. Honeybees are complex, idiosyncratic, and incredible creatures, and for most of our lives we barely noticed them. We agreed to meet up at the next class. But that week we both were becoming preoccupied by bees, exchanging videos of bees through emails, reading children's books about bees, compiling lists of questions about bees' habits, and worrying about how we might have become interested too late—bees were in trouble, "disordered," and missing.

Through this book, we want to add to the buzz by collaborating on some kind of seismic activity between phenomena. The buzz, as a conceptual reverberation or an echo, is where the sound bounces back and forth between objects. It speaks to the exchange between organisms that is often obscured by humans and humanism and to the tendency toward preoccupation with the needs of our own species. As we got closer to bees, the hypnotic sounds they made rang in our ears even though we couldn't decipher their language. We didn't decode their buzz, but we could sometimes feel the bees' intense humming in our own bodies, so that everything around us—all objects, animate or inanimate—were drowned out. The bees got our attention and they made us think. We first encountered them as "fun" and interesting research subjects, and eventually we came to reconsider them as living organisms and co-species, as we will discuss in the pages ahead. At times, thinking about our multifaceted relationships with bees led us to question our own actions and choices, the problems that stem from our humanness—a positionality that we were not taught to challenge or even ponder. In this book, we continuously grapple with the human practice of seeing bees as signifying something else, through metaphor, and approach an ontological reckoning with the insect.

Like all living organisms, we grew up in a multispecies world but received the message that humans were unquestionably at the top of the so-called animal kingdom. In this way, humans are naturally off the hook: we can interact with certain species on our terms and conditions, avoiding or exterminating certain undesirable creatures or working to save others we deem majestic or useful. Humans have the ability to choose the "it" animal, the animal that we will rally to support and admire. There are certain nonhuman species that demand and receive more human attention, for example, since the 1970s environmental groups have championed saving the whales—a species most of us have little daily contact with. And now at the start of the 21st century, humans are concerned with saving bees, as we learn to look more closely at the ways in which these seemingly banal insects so directly impact our human way of life.

A box of clamoring, buzzing bees will demand your attention yet is a temporal entity. Hives are temporary homes because ultimately bees will come and go as they choose to, whether they swarm in search of more space or simply disappear altogether for reasons that elude us. Bees cannot be wholly contained, not in a hive or in theory, at least not permanently.

For the most part, people usually come face to face with bees themselves in rather ordinary albeit intimate ways. Most of us have witnessed a bee landing on a flower or swatted at one that accidentally made its way into our personal space uninvited. Bees can be enthralling and familiar but at the same time may be startling and alarming. They have a capacity to lure us in closer so we can observe the "beauty of nature" and to send us racing with the fear of getting stung. They can swarm loudly in groups of thousands but can also hum while placidly collecting pollen on spring blooms. The sounds of bees have inspired musical recordings, and bee sonic variations have been sampled in songs as a form of instrumentation. Sensitive beekeepers listen carefully to the buzz of their hives, assessing their moods and temperament with respect to pitch and tone.

The buzzing sound is generated not from their heads but from the waving movement of their two pairs of wings. The bees' anatomy consists of three segments; the head, the thorax, and the abdomen (see figure 2.2).

The head of a bee contains two sets of eyes, an antenna, and a long tongue. The large compound eyes are used for distance vision, and simple eyes (ocelli) at the tops of their heads allow them to register light. The thorax, or midsection, contains three sets of legs and two sets of wings. The third pair of legs, or hind legs, possess pollen baskets (corbicula) on each leg that collect pollen as they transport it from flowers to the hive. Honeybees have a larger set of front wings and smaller ones toward the back ends of their bodies, which they beat between 200–230 times per second. As the photographer Rose-Lynn Fisher illustrates in her extraordinary magnified images of bees, the top edge of each hind wing has "hooks" called *hamuli* that "catch on a fold of the bottom

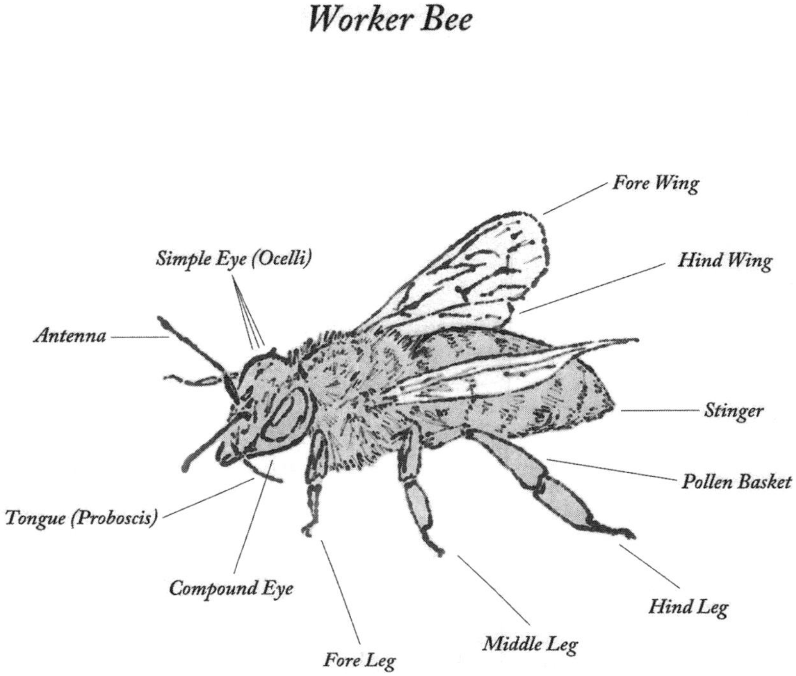

Figure 2.2. Illustration of a worker bee. (Illustration by Mike Schreiber)

edge of the forewing."[4] When bees fly, the two sets of wings temporarily fasten and separate. With magnifying photographic technology, Fisher reveals this movement in a photo titled "wing seam 160x," which is described simply as "the wings hinged together."[5] The minute detail of the wing seam resembles pieces of fabric sewn together with sturdy thread; the image looks like the hem of an exquisitely tailored garment. The human eye cannot see this microscopic fusion, but through this process the fanning is made audible.

The different meanings of the word *buzz* are relevant for understanding the exchanges and intersections between humans and bees. Buzz is not just applicable within this context as a sound, but also as a physical sensation, an emotional state. We have heard the colloquial phrase "catch a buzz," and most of us can relate to the feeling of being high or elated, whether through drugs, exercise, sex, or our body's natural endorphins. *Buzz* connotes the presence of energy, too, as in the cliché of a room buzzing with excitement. Bees embody some of these forms and meanings of buzz. As we discuss, they have the ability to not only get beekeepers and environmentalists excited, but also make the ostensibly uninterested curious and concerned.

Of course, if bees do indeed stop buzzing it is a tangible reminder of human fallibility and larger environmental changes. They embody life literally and metaphorically, and their disappearance also suggests the fragility of life and a potential disruption in how we live. Humans are not directly interconnected to any other insect in such a profound manner; through pollination and the consumption of honey, bees become part of our bodies. There is a tangible ecological, economic, agricultural, and physical link. For many beekeepers, this relationship is clearly emotional and intimate, an interspecies exchange of life and labor.

Urban Beekeeping

Even though bees are a new cause and sometimes labeled as trendy pets, beekeeping has not always been a welcomed practice in the city. Bees have been considered as territorial invaders, and human-bee relationships, therefore, have also caused controversy within urban municipalities. In March 2010 the New York City Department of Health voted favorably regarding lifting a decade-long ban on beekeeping. Interestingly, beekeeping was more common in the city in the first few decades of the 20th century, with notable hives kept on top of Radio City Music Hall and the American Museum of Natural History. The practice was officially outlawed in 1999 under Mayor Rudolph Giuliani, "when honeybees were included on a health code list of more than 100 wild animals that New Yorkers could not keep, including vultures, iguanas, ferrets and even whales: they were all potential menaces."[6] Although keeping hives had been illegal since then, hundreds of people continued to defy the law. Urban beekeepers became outlaws of sorts, though the city didn't devote much time and energy to ferreting out hives—only a handful were ticketed during the ban. It was reported that in June 2009, "the city had received 49 complaints," and "officials made nine inspections and issued four summonses."[7] Tickets for beekeeping were comparable to being ticketed for jaywalking, both minor offenses that slip under the radar. However, the minor offense can seep into human consciousness enough so that some may wonder if this is the time that they might get caught.

And yet the bees are here. Due to a revived interest in urban farming, locavore food movements, green consumerism, and a demand for gourmet boutique honey, bees occupy an increasingly visible role in the ecological and cultural life of New York and other major American cities. Bees now thrive in some urban environments, even as entire hives of bees have been disappearing in vast numbers because of a syndrome referred to as Colony Collapse Disorder (CCD). As discussed in the next chapter, CCD accounts for the unexpected loss of 30 to 90 percent of hives in the United States. Because of the wide media attention to CCD since 2006, humans have started to pay closer attention to bees. We have come to appreciate their labor, either indirectly by way of pollination or through purchasing products made from their pollen, wax, and honey. But humans not only consume the fruits of the bee for the sake of health, environmental sustainability, or culinary fads, they also establish and nurture their own bee colonies. Cultivating your own honey involves more commitment than buying it at a food co-op. We have encountered so many people who want to learn more about the bees, get closer to them, and even help them. A young musician who lives in Williamsburg, Brooklyn, recently asked if we knew of a beekeeper who needed helpers because he wanted to "save the bees!" This is difficult to do without any knowledge of beekeeping, but his interest was earnest.

Although we tend to view the urban environment as a peopled place, animals are everywhere—from rats that run too close to our legs while waiting for the subway, to Chihuahuas dressed in pink shirts peering out from Louis Vuitton bags. Cities are centers of human life and are also dense places of intersection for humans and nonhumans. From a sociological perspective place is not simply about space, or an area defined by more technical, territorial, and bounded terms. As the sociologist Christopher Mele asserts, "Place is space filled up by people, practices, objects, and representations."[8] Place is made meaningful over time because it is embodied, lived in. Through memory and experience it becomes a human landscape evoking feelings, emotions, and nostalgia. Its inhabitants give a community a sense of place, and in our fieldwork urban places were made meaningful in part through the presence of unlikely insect neighbors. Following the geographer Jennifer Wolch, we bring the bee into the urban landscape for intellectual, ecological, and moral reasons "to re-imagine the *anima urbis*—the breath, life, soul and spirit of the city—as embodied in its animal life."[9] In this book, we place bees within cities and other human environments, but we don't displace the bee itself. We attentively recognize the bee's existence (as best we can from a human perspective), as a living organism in its own right, rather than as a metaphor of productivity exclusively or as a species we have come to rely on.

We weren't quite prepared to find bees in locations that offered little evidence of nature in the form of trees, plants, grass, and other living organisms. Humans make homes for them wherever they are able, and with a shortage of real-estate and green space, bees, like humans, make do in areas that are less than what is conventionally thought to be scenic. As shown in figure 2.3, Selena, one of our informants, was sugaring her bees on a rainy rooftop in Brooklyn where they industriously lived amid brick, metal, concrete, tar paper, plastic, dirt, and garbage.

Sugaring is a form of pest protection, and it's an organic and simple pesticide method that presumably keeps bees safer and healthier. Even though some beekeepers attempt to control and protect their bees through careful monitoring and protection, the immediate environment can affect them in particular ways. For example, in spring 2010 in the Red Hook section of Brooklyn, an increasingly gentrified neighborhood, bees started glowing red from gorging on maraschino cherry juice made at a nearby factory. Their stomachs, and in turn their honey, was affected by the bright red dye, alarming local beekeepers who were concerned that bees should be surviving on plants rather than sweetened syrup. The journalist Susan Dominus posed an interesting question about the preferences of these urbanized bees:

> It seems natural, by now, for humans to prefer the unnatural, as if we ourselves had been genetically modified to choose artificially flavored strawberry candy over strawberries, or crunchy orange "cheese" puffs over a piece of actual cheese. But when bees make the same choice, it feels like a betrayal to our sense of how nature should work. Shouldn't they know better? Or, perhaps, not know enough to know better?[10]

Figure 2.3. Selena sugaring bees on the roof of a synagogue in Brooklyn. (Photo credit: Lisa Jean Moore)

While bees may not "know better" to steer clear of food saturated in red Dye No. 40 (much like humans), the cherry bees of Brooklyn are proof that insects adapt to and are directly affected by their surroundings. For better or worse, the city presents obstacles and opportunities for all of its inhabitants. And as honey goes, according to our informants, the Red Hook red honey was tasteless and deeply disturbing in its shocking redness.

As the song declares, "If I can make it there, I'll make it anywhere." Thriving and surviving in New York City has been described as the ultimate test of human resilience. At a minimum, it's a barometer of some combination of moxie, street smarts, and stamina. Throughout our fieldwork we tried to resist the temptation to anthropomorphize the city as a hive, with apartments as cells, people as workers foraging and returning, always kinetic, buzzing, humming twenty-four hours, every day, nonstop. Such observations seemed superficial and obvious. However, there is something to the urban survival narrative that links humans and bees. When bees live within the same crowded, toxic, noisy, imperfect conditions, maybe it can be read as some kind of omen to beekeepers and the rest of us. If tiny flying insects can survive within landscapes of overconsumption, pollution, and gentrification with limited resources, maybe there is hope for humans. In the book *The Queen Must Die*, William Longgood underscores the significance of bees and our interspecies connection:

> The bee is domesticated but not tamed. She has not recognized man as her master; he subdues, manipulates and beguiles her into working for him, but the bee remains what she has always been, part of nature, a part of ourselves forever lost, part of the joy and sadness in the long march from the unknown beginning toward an unknown ending, like man himself, part of the great inconclusive experiment called life.[11]

Urban beekeeping demands energy and knowledge—you can't simply go to a pet store to buy bees and a hive. The work of beekeeping is dictated not only by landscape and availability but also by seasons. It is a cyclical practice set in motion by the ebbs and flows of nature. During the winter, the bees use survival strategies called "overwintering" to make it through the cold when there are limited opportunities for pollination; bees will weather the chill by vibrating and beating their wings to create heat. They don't technically hibernate because they never sleep. Bees are forced to stay inside, but they are, in essence, active captives. The bees' outdoor season begins in spring when the temperature is warm enough for them to emerge from their hives, typically when a few consecutive days of about 60 degrees Fahrenheit or above occur. For the beekeepers, the season begins months before spring arrives, as people either arrange to buy bees (for between $35 and $100, depending on the presence of a queen) from beekeepers who sell them professionally or to inherit them through local beekeeping connections. Beekeepers obtain what is called a "package," beekeeper terminology for three pounds of bees and a queen. The package is also referred to simply as a "nuc," or nucleus colony, which is a smaller colony of bees centered around a queen that has been bred from larger colonies. The bees are installed into hive boxes and are sometimes fed pollen patties to help them get established over early springtime. Beekeepers then check the bees periodically to ensure that the queen is healthy and laying brood (eggs from their embryonic to larval states). The brood chambers, which are the hexagonal shaped cells that structure the hives, are mostly filled with worker larvae. But beyond the beekeepers' vigilant eyes, honeybees themselves know what is needed to make the hive work. They are "connected by the network of their shared environment" and "achieve an enviable harmony of labor without supervision."[12] We work with and for bees, but they don't necessarily require our assistance.

In densely populated vertical cities like New York, backyards are a luxury (one that is often shared) and not all neighbors are keen at the prospect of inviting tens of thousands of bees into collective outdoor spaces. While some people are able to set up hives in their backyards (we met two who did so during three years of research), they are more commonly located in community urban gardens, apartment rooftops (as pictured in figure 2.4), or attached decks, wherever space and access allows. Beekeepers do not have to check their hives every day, and they don't have to commit to regular walks and feeding as is the case with nurturing dogs and cats.

But keepers do need equipment (like basic hive tools) and protective gear, which can take the form of a full beekeeping suit; a veil, hat, and gloves; or some variation. Some beekeepers are comfortable wearing pants and long shirts, and we even met someone who purposefully tends to his bees barefoot. Beekeeping styles and philosophies vary widely, often reflecting the personalities of each beekeeper. After idiosyncratic prepping and preparing, people typically have to commute to the bees. For some, this can mean a forty-five minute drive or subway ride, trudging up five flights of stairs, braving precarious metal fire-escape ladders, and other inconvenient physical obstacles, all with equipment in tow. Throughout spring and summer, beekeepers visit their hives once a week or more, and while the process may become routinized, a hive check is more of an event, sometimes bordering on spectacle. It's not like feeding a goldfish. When a box of bees is opened, everyone within a few feet of the din and frenetic movement takes a step back.

Figure 2.4. Hive boxes on a Crown Heights rooftop. (Photo credit: Lisa Jean Moore)

Model Insects

Bees are technically insects, but they are not bugs in the pejorative sense of the word. Unlike ants, mosquitoes, roaches, and spiders, which tend to fall under the spectrum of bugs and pests, bees both exemplify and transcend these designations. Our personal connections with bees and their cultural significance are appreciably more complicated. Most see insects as a nuisance or as a potential threat due to the possibility of a sting or infestation. Certainly, there are people who have pet tarantulas and ant farms, but in these cases they exist behind glass where they can be observed and displayed safely. These pests turned pets function as exotic insect others, something akin to a curiosity cabinet with the focus on spectatorship.

Our focus in *Buzz* is mainly on honeybees, one type within 20,000 species of bees, each with its own idiosyncratic characteristics and behaviors. Typically, only experienced beekeepers and entomologists can quickly visually identify a carpenter bee from a mason bee. For the uninitiated, honeybees, which are very docile, can also be easily misidentified for other flying insects that are similar in size, coloring, and appearance. Hornets, yellow jackets, and wasps, which are more aggressive and territorial, are sometimes confused for honeybees. However, honeybees are the most anthropomorphized and the ones we are most familiar with—soon followed by the bumblebee, which makes honey but in much smaller amounts. When people keep honeybees while the hive boxes are stationary, the colony itself is not contained or stationary. The bees come and go as they please. In this way, bees do not make good pets per se. For example, honeybees can visit between fifty and one hundred flowers a day on their nectar and pollen collection trips, and a hive as a whole will visit about two million flowers in order to make one pound of honey. Without mobility and autonomy, the hive will perish.

Honeybees are social insects that live in colonies and maintain their survival through harvesting nectar and pollen from flowering plants and, in New York City most especially, from tree blooms. Flying from blossom to blossom, they enable pollination or fertilization and sexual reproduction of plants and trees. Honeybees convert the nectar to honey through enzymes in their stomachs. Honey, referred to humorously as "bee barf" by an informant who is a seasoned beekeeper and enjoys lecturing to middle schoolers, is used for sustenance and to raise brood (their offspring), and the pollen provides protein for the bees' diet. While some species of bees can survive as individuals, honeybees cannot. They must be part of a colony, what the entomologist Thomas Seeley describes as a "harmonious society, wherein tens of thousands of worker bees, through enlightened self-interest, cooperate to serve a . . . common good."[13]

In many ways, honeybee colonies are prototypes of sustainability and its members are experts in arcology—a form of design that merges architecture and ecology in structures made to accommodate densely populated areas. They design living spaces of hexagonal cells that are jam-packed and multipurposed—used continuously for reproduction and storing honey. They create all they need from their environments and grow their own homes through producing beeswax made from secretions from their abdomen. Honeybees are praised for their efficiency and interdependency, are adept architects, and have become a species that humans rely on. For these reasons, bees are model insects. We have continuously looked to them as an example of how to produce and to prosper as a species, without draining or harming the planet's ecosystems. As the philosopher Freya Mathews has written, bees are integral to the planet's "inexhaustible regeneration of life" through pollination, one of the "great metabolic processes of the earth" in addition to photosynthesis and thermal and atmospheric regulation.[14] And yet, as we argue, it is only when they go missing that bees began to tangibly appear to us.[15]

Bees' economic and agricultural utility to humans is quite considerable, and we do not rely on any other insect as much as the bee. They pollinate flowers and crops that provide our food supply, operating as nature's invisible labor force. The annual value of honeybee pollination to U.S. agriculture is estimated at more than $15 billion.[16] They also produce honey and wax so they have a direct and obvious use value that is not inherent to other insect species, save for silkworms. However, in comparison, while silk is a valuable commodity, produced primarily in China and India, it accounts for less than 1 percent of global textile production.[17] Silk is a luxury product and not a part of everyday consumption. On average, every American consumes nearly one pound of honey per year.[18] Honey also shows up on our tables in a more clandestine way—as an ingredient used by industrial bakeries and cereal and health food manufacturers. Bees are the only insects that produce food that is eaten by humans. So the tangible work of bees is often veiled by plastic honey-bear containers or tubes of organic beeswax lip balm, as we tend to take it for granted that this is just what bees do *naturally*. Humans are usually pleased to coexist with (and profit from) this insect, as long as it stays out of our paths and behaves predictably. When bees fly into our space, it is alarming, triggering fear and panic. Humans tend to prefer bees that are in some way contained—accessible, predictable, and at a safe distance.

Beyond humans, bees also provide food for species like bears, badgers, skunks, and raccoons. Rodents and other insects are lured by the honey and the pungent aroma the hive gives off—cockroaches, hornets, wasps, ants, and mice will get into hives and eat the wax, honey, and even the honeybees themselves. We talked to one beekeeper who found a mouse in one of his hives that was completely entombed in wax. We joked that it was like a little mummy or sarcophagus and he told us that Alexander the Great was enshrined forever in a reservoir of honey. We learned that the Egyptians, Assyrians, and Hindus used beeswax and honey as a type of ritualistic embalming fluid. In *Robbing the Bees*, the beekeeper Holly Bishop writes, "Egyptians took lessons in embalming from the bees, wrapping bodies in wax-dipped linen bandages to preserve them for the afterlife. Ears and nostrils were often packed with wax and resins before the body and its removed organs were placed in a coffin or series of coffins sealed with beeswax."[19] Beyond the functionality of honey and wax to prepare corpses, past civilizations found other uses for bees. Beeswax has been used to buff, preserve, and waterproof a vast range of objects from leather and fruit to the surfaces of boats.[20] The Egyptians derived benefit from bee venom medicinally and symbolically—bees were called the tears of the Sun God Ra, and they represented birth, death, and resurrection. Both humans and animals have found creative and purposeful ways to make use of bees and their products for thousands of years.

Animal Studies

Like bees, humans are social creatures who would likely not survive as solitary beings. On a basic emotional and psychological level, we require intimate and physical connections with significant others in order to, in a sense, "feel" and be human. We also reach out to other nonhuman species for various comforts, whether through a pet cat that purrs at our touch or the utility of a chicken laying an egg that becomes our breakfast. Urban beekeepers who establish hives shed light upon how and why we continue to seek connections with other species, to literally make contact with insects. The word contact borrows from the feminist and biologist Donna Haraway's notion of "contact zones," a conceptual term used to describe the entanglements between species who do not share languages but are otherwise co-present and co-mingling organisms.[21] Haraway's work questions human-ranking systems, calling attention to how "the odd singular words *human* and *animal* are so lamentably common in scientific and popular idioms and so rooted in Western philosophical premises and hierarchical chains of being."[22] The notion of contact zones cleaves at the long-standing and often unchallenged hierarchy of man (as a higher being) and animal (as a lesser creature).

This book itself is a contact zone, a scholarly effort to describe how humans come to know bees and how bees react to different forms of human action—from altruistic attempts to save them to environmental practices that unwittingly harm them. We are also hoping to take part in conversations that diverse scholars, including entomologists, social theorists, and expert beekeepers, have been having about these issues. But because we are interested in the collisions and communications between humans and nonhumans we also draw on a relatively new field of academic research that has come to be known as "animal studies." Emerging as an area of inquiry in the 1970s and 1980s, animal studies brings together scholars from many fields including anthropology, sociology, cinema studies, history, psychology, and feminist and queer theory (among others), and it has increasingly explored the complex relationship between humans and nonhumans. Among the many issues that these scholars have addressed, the most salient for us is the animal/human binary that has been considered in relation to other binary systems such as masculine/feminine, gay/straight, and nature/culture. As the sociologist Jen Wrye succinctly states, "The animal/human divide remains a key feature of modern life."[23]

Animal studies was partially developed out of the Animal Liberation Movement of the 1970s that emphasized humans' moral relationship to animals and that questioned the ethics of human/animal interactions. Some working within animal studies focus not only on how we conceptualize nonhumans but also on what we do to certain animals as a matter of course, like cage, slaughter, and eat. The philosopher Peter Singer's *Animal Liberation* (1975) is arguably one of the most influential texts on the ethical treatment of animals.[24] Graphic descriptions of animal testing—from psychological experiments in which monkeys are raped to chemicals dropped into the eyes of conscious rabbits—provide the foundation for his main argument: animals are sentient beings affected by torture and physical pain and are capable of suffering. Singer reasons that the ability to suffer is what confers basic rights for all living things, regardless of species. Importantly, some philosophers and animal rights activists assert that evidence of animal intelligence, rather than the ability to feel pleasure or pain, should be the measurement for sentience.[25] In arguing for animal rights as evidenced by suffering, Singer challenged human-centered "speciesism," or the assumption that one species has more value and, thus, the right to act on other species in their own interest. Ethical veganism, as a philosophically driven dietary practice, is sometimes grounded in the ideas of *Animal Liberation*, a work that has resonated beyond the academy.

Animal studies scholars have mainly focused on domestic animals and pets, rather than on species we consider to be exotic or wild. The absence of scholarly work on alligators or zebras is perhaps more pragmatic than discriminatory, as humans typically share their daily lives with dogs and cats. There has been much attention paid to the psychosocial aspects of companion animals as being central to human well-being. Even though pets may be an economic drain and consume much of our time, studies suggest that they are worth the effort to humans from an emotional standpoint. For example, the anthropologist Joel Savishinsky has researched the therapeutic benefits of pets within institutionalized settings like nursing homes. According to Savishinsky, pets function as "transitional objects" "through which patients can overcome insecurity, create ego boundaries, and go on to develop a widening circle of warmth, approval, and social interaction."[26] Similarly, the psychologist Judith Siegal asserts that companion animals provide humans with both comfort and stress reduction.[27] Scholars have also explored how pets can function as a type of cultural capital[28] and as an extension of the owner's sense of self-identity.[29]

We love our pets, much like we love human significant others: we name them, celebrate their birthdays, and buy them special gifts. But what is important to note is that humans do not treat all animals within the same species as beloved companion animals. Humans can "regard the same animal as both a companion and an object" as is the case with hunting dogs or those bred to fight.[30] As the sociologists Arnold Arluke and Clinton Sanders point out, "Ambiguous perceptions and ambivalent emotions are central to the forms of relationships between humans and nonhuman animals."[31] Animals can be read as functional objects, like cows or horses, while others that are imbued with pet status are believed to be sentient creatures with personalities. Arluke and Sanders argue that the animal/human divide is clearly not a simple dichotomy, as our definition of animal (pet or otherwise) itself exists on a continuum. Our complicated relationship with bees, as both model insect and feared stinging invader, speaks to the murkiness of animal/human intersections.

The terrain of animal studies has become a rather pioneering field in academia, but there are some scholars who advocate beyond theorizing about animals or describing the conditions under which we come to know them. There is a split in the field of animal studies whereby there are those who do mainstream animal studies and those who practice critical animal studies. As discussed by Steven Best, a philosopher and animal rights activist, trepidation concerning the surge of scholarship on nonhuman animals is growing:

> For academics whose commitment to animals is strictly abstract and theoretical, nothing more than an interesting topic of research and form of academic capital, there is no contradiction here. But for anyone who understands the real, concrete suffering of animals and the logical consequences (i.e., veganism and animal liberation) of valuing them as living beings rather than as signs, referents, texts, and publications, the contradiction of speciesists working in the field of animal studies is startling. . . . After all, it's fun, interesting, the new wave, "progressive," and the scholar who begins work in this field might get some new publications, make new contacts, kick-start an incipient career, or revivify a flagging vocation. Thus, one finds carnivores, pro-vivisectionists, and garden-variety speciesists operating in an academic terrain where a considerable number of theorists view animals as historical referents and abstract objects of research, rather than giving urgent attention to those beings who live and suffer now, to the thousands of species teetering on the brink of extinction, and to the profound obligations we have as scholars to dramatically highlight these problems and to take aggressive action to protect and liberate present and future generations of nonhuman animals.[32]

Best is wary of how academics sometimes objectify animals as mere things to study from a distance. We ponder them but too easily set aside the fact that they are living organisms that are affected by human desires and systems of production and consumption (like monocropping or the beef industry). Saying something about the maltreatment of animals is different from acting to eradicate it. Thus, he urges a commitment to end animal suffering through pursuing animal liberation, animal activism, and veganism.

We are conscious of how our work straddles the divide between mainstream and critical animal studies. We might be characterized as

Best's "fun scholars" who understand the bee as an object manipulated by humans and an insect produced in multiple texts for deconstruction. Or we could be considered as scholars who see the bee as an organism that may well be suffering due to human interventions in the biosphere, and consider this suffering as in need of an immediate remedy. Taking into account the imbalance of power in our nonreciprocal relationships with our insect subjects brings up questions of interspecies ethics and exploitation. As researchers entering sites of human-bee interaction, our access to bees is facilitated through their human caretakers; thus we are already at a distance from the bees. We come to know our nonhuman subjects in large part through their keepers; we were never alone with bees while in the field.

This inherent disconnect and imbalance of power is addressed in the field of critical animal studies. While aligned with animal studies as an interdisciplinary approach to investigating human/animal relationships, critical animal studies advances two frames of critique: "first, a critique against animal studies itself and its often accompanying detachment from the actual life conditions of most animals; second, a critical theory approach, broadly defined, to human/animal relations, with close attention to concrete forces of power and resistance."[33] Because critical animal studies is allied with animal liberation and activism, these scholars advocate the merging of theory and practice. Importantly, academic knowledge itself is critically interrogated, emphasizing the politics of how certain frameworks become taken for granted. There is also a commitment to examine sites of oppression where animals and humans intersect. With this in mind, we work to interpret all of our fieldwork and analyses with particular attention to the everyday life conditions of the bee.

We are deeply motivated to study bees as a species that are deployed in multiple nonconsensual arenas, as is the case with industrial pollination where bees are loaded en masse in trucks and driven across the country to pollinate almonds and other crops. With the advent of massive hive disappearances in the wake of CCD, more people are interested in what may be the cause of the bees' suffering. In this light, we not only describe what various people do to and with bees but also suggest ways in which we can alter our behaviors in order to create less adverse conditions for them.

Just as critical animal studies challenges a human-centered approach to theory and research, the area of posthuman studies is also fruitful in helping us re-position ourselves in the field, theoretically and methodologically. Posthuman studies have grown out of questions about the construction of the "human"—an increasingly difficult object to define due to technological enhancements, such as steroids and prosthetics, and projects of hybridization, such as xenotransplantation (the transplantation of different species organs into a body) and genetic manipulation. It is increasingly difficult to simply define what it means to be human without considering our species as enmeshed with nonhuman animals, bacteria, medical devices, and aids. Fields such as science and technology studies, feminism, philosophy, and animal studies advance a posthuman philosophy. Like many academics before us, we were trained in the traditions of American humanism. In general, within the academy, traditional humanism is predicated on the idea of humanity or humankind as a fact or starting point for scholarly inquiry. Humanism emphasizes that ethical life practices should aspire to the greater good of humanity, leaving no space for other species or even for human/technology hybrids, something we are likely to see more of in the future. Posthumanism, a more recent innovation, argues for "a deconstruction of symbolic, discursive, institutional, and material arrangements that produce the category of human as something unique, distinct, and at the center of the world."[34] As human ethnographers, we constantly grapple with speaking for and about a nonhuman species. Yet, as the bee's interlocutors we assiduously attempt to de-center ourselves, and yet we are aware of how our training in humanism may unconsciously affect our ways of seeing.

Studying Bees: Insect/Human Ethnographies

Our work delves into human and nonhuman worlds, which raises questions about the limits of ethnography. *Ethno* means people or folk, and any ethnography attempts to understand and explain a people and their culture through writing about them. However, we interacted with both beekeepers and bees. Even though bees do not speak, we wanted to be able to engage them as "informants" while at the same time we were uncertain as to whether that would be possible. We have learned a great deal from consorting with the bees themselves. Following the work of the anthropologist Eduardo Kohn, we seek to expand the reach of ethnography, to focus "not just on humans or only on animals but on how humans and animals interact."[35] Reflexively, we are performing an "api-ethnography" that considers bees as cultured beings that traffic between worlds of the hive and of the urban landscape. We situate our work as part of the growing contributions in multispecies ethnography, located at the intersections of environmental studies, science and technology studies, and animal studies that are working to bring understudied organisms to light.[36] Multispecies ethnography is a new genre and mode of anthropological research seeking to bring "organisms whose lives and deaths are linked to human social worlds" closer into focus as living co-constitutive subjects, rather than simply relegating them to "part of the landscape, as food for humans, (or) as symbols."[37]

As with many ethnographies, "research" is told through stories where characters (most commonly humans) are presented in rich descriptive paragraphs seasoned with the characters' own words. These words, excerpts from interview transcripts, are then interpreted and provided as evidence of larger theoretical claims. They become "data," speaking of not only singular experiences but also of ethnographic writing and academic knowledge. We are cautiously confident about our ability to interview and understand the urban beekeeper as a member of our own tribe. Understanding the subjective experience of bees is a more complicated endeavor. Without a shared language, we could not reciprocally communicate with our "other" research subjects who were almost always present. We observed and interacted with tens of thousands of bees, collecting copious data that attempt to describe their lives. Yet now we must interpret bees' behaviors in imperfect human terms. Ethnographic training has not adequately prepared us to speak for and about a nonhuman species.

Social scientific studies of insects are not unheard of. Indeed, the sociobiologist and Pulitzer Prize winner Edward O. Wilson has dedicated much of his research to understanding the life of another social insect, the ant.[38] Wilson famously likened ant behavior to a form of socialism where self-sacrifice for the good of the colony is commonly practiced. As a biologist, Wilson also "discovered" how ants communicate through the use of pheromones. So captivated by the species,

Wilson has also written a novel entitled *Anthill*, which features battling ant colonies and human land prospecting. Controversy over Wilson's sociobiological ideas is well documented, as scholars question the theory that there are evolutionary links between insect and human behavior.[39] Importantly, we are not using the bee to claim that there is an evolutionary or biological linkage to human behavior. We interpret how humans culturally deploy the bee to inadvertently or even deliberately make statements about human nature.

In some of our vignettes, human storytellers share their own bee narratives—they are implicated in the stories we tell about them. Bees are perhaps not "silent witnesses" but buzzing witnesses to humans, whereby children and adults are narrating the bees' behavior through the somatic clues (the buzz, the sting, the smell, the productivity). As sociologists, we are most comfortable with this type of analysis, revealing and criticizing the anthropomorphizing that humans (including ourselves) do to bees.

Transcripts and photographs become the familiar terrain for us to dissect and examine as we search for the lurking larger narrative arcs and relationships to human order and disorder. We feel safer in our analysis when we are examining bees as an object in relation to other objects; everything is brought into being, including bees, through their relationships.[40] Perhaps one could think of CCD in terms of bees experiencing a falling away or becoming out of sync with human/technological agents, and a falling back to something else, not nature really, but some hive-driven protective mechanism that preserves production in some future generation. Or maybe the bees are falling away decisively from their relation with a human-prescribed diet. As one holistic beekeeper told us, "Bees just want to be bees," suggesting that perhaps it is humans who should fall away from them.

Critical animal studies researchers suggest that we become advocates for the animal and set aside our human impulses, and yet as ethnographers we are limited with few tools to inhabit that space of beeness. It is decidedly more difficult to interpret these nonhuman actors. We don't speak their language, share their culture, or engage in mutually negotiated intimate acts. Because we can't have direct relationships with the bees, we are engaging more in practices of circulating reference. Bees certainly exist outside of human consciousness, but how can humans know bees without being limited by their humanism—their speciesism? We examine the intimacies we have attempted to establish with bees—but there is something missing in the examination as we can apprehend bees only through our limited senses. What we smell, taste, hear, and feel, in addition to what we *think* about bees, is filtered, diluted by humanness.

And yet, perhaps, bees aren't so foreign to us. Maybe we're inextricable from one another, as suggested by the philosopher Timothy Morton's *Ecological Thought*.[41] Morton asserts that all forms of life are connected within one vast, entwined web and that no being, idea, or object can exist independently of our shared environmental entanglements. In a very real sense we all embody bees throughout the course of our lives because they are an integral part of our everyday diets. Bees produce much of the food we consume, whether through pollination or honey production. The labor of bees constitutes us physically. We also consider bees metaphorically as highly productive model insects that we can learn from. For these reasons humans should ensure that they and their descendants make it through the anthropogenic mass extinction. After all, other economically useful species are threatened and have gone extinct, but they haven't been nearly so embedded into human culture. Bees have inspired poetry, literature, art, and music in addition to countless metaphors such as being "busy as a bee." As one colleague reminded us, the usage of honey as a term of endearment alone is noteworthy as nobody has ever called their beloved "stellar's sea cow lamp oil."

We are not bee whisperers and we are suspicious of those who claim to be—rather we tentatively suggest we are experts on describing certain human beings and their behaviors and actions. For us, and for other humans, the bee is a cultural artifact that has its own historical and temporal social location; the bee does things to cultural life just as the bee does exist as a real and material insect. We are human interlocutors interrogating our own and others' nonconsensual use of the bees at this particular moment in contemporary human life. We want to better understand what bees might need to live more productively; however, our quest for what the sociologist Max Weber referred to as *verstehen*, or empathetic understanding of our research subjects, is deeply limited and suspect. In our project, we engage in this murkiness and clumsiness

Figure 2.5. Bees in a top bar frame. (Photo credit: Lisa Jean Moore)

regarding the cultural and discursive idea of the bee, humans' own rendering and use of them, and the actual bee as its own being.

Engaging with the work of the physicist Karen Barad, our challenge has been to allow bees to emerge in "intra-actions"[42] where "the subject and object do not pre-exist as such, but emerge through intra-actions." Bee and humans are entangled entities that mutually constitute one another. As sociologists, we need to interrupt our tendency to think of bees as the object of study. Similarly, we have to resist thinking of ourselves or the beekeepers as static, bounded, and permanently fixed entities. Instead we need to see all actors—ourselves, bees, the beekeepers, and other objects—as bodies that are in the world and whose boundaries are created through intraspecies entanglements and conflicts.

* * *

In what follows, we describe in detail the ways we engaged with urban beekeepers, bees, and their ecologies over the past three years. Standard for most qualitative research studies, we feel it is important that we provide the methodological activities and decisions that frame our burgeoning epistemologies of bees and humans—their relationships and interactions. It is more the ontology of the bee, as both entangled with humans and as beings/things in their own right, that we struggle to reckon.

As qualitative and reflexive sociologists, we have to consider that our first encounters with the specter of the "bee" have framed our relationship with the species for our entire lives. As children of the 1970s, so-called killer bees were a prominent threat to our safety and security, and this threat was profitably dramatized in popular culture. A sample of films that came out in quick succession include *The Deadly Bees* (1967), *Killer Bees* (1974), *The Savage Bees* (1976), *The Swarm* (1978), *The Bees* (1978), and *Terror Out of the Sky: The Revenge of the Savage Bees* (1979). We vividly remember many of these films, including how they made us feel about insects. These films illustrate how the cultural meanings of bees can change over time and how our own relationships to these insects can shift—from seeing them as frightening predators as children, to becoming comfortable enough to do a hive inspection with no veil or protective equipment. In our lifetimes, however, bees have occupied the spaces of being threatening and, more recently, being threatened. While other animals are scary (e.g., bears, sharks, rabid dogs, piranhas), our chances of being assaulted by them are very slim. Bees are more likely to fly into our personal space, which can lead to a startling painful exchange. Unlike our relationships with many other insects and animals, there is a physical interaction between humans and bees: an exchange of fluids, a co-mingling of bee venom and human flesh and blood. Getting stung by a bee is a painful albeit temporary experience for humans, but from the perspective of the bee it signals finitude—death, because the bee's body is then ripped apart and it soon dies. The sting alludes to emotional exchange—one of fear, but also of feelings of caring, and responsibility toward living organisms. With this in mind, we explore the intimacies (literal penetrations) between humans and bees, and the ways in which the lives and, in particular, the bodies of species are enmeshed.

The primary way in which many of us learn about bees is the mass media. As a public sphere, it reflects political, economic, and cultural ideologies.[43] We analyze websites, as well as print media (newspapers, magazines, children's books) documentary films, and advertising. Our aim is to understand how the bee is mediated by culture and inscribed into the public imagination. However, our ethnographic analysis goes beyond cultural and media studies. We are particularly interested in exploring how bees are being used within the burgeoning environmental movement, green consumerism, and ecopolitical discourses. In light of innovative conversations emerging within multispecies ethnography, we elucidate "how a multitude of organisms' livelihoods shape and are shaped by political, economic and cultural forces."[44]

We examine relationships between humans and honeybees within the five boroughs of New York City and a few Italian villages; between novice and expert beekeepers; among queens, drones, and workers; and also through historical migrations and multinational flows of capital. Our theme of buzz is central to each chapter but in slightly different tones. In chapter 3 we begin with the ecological buzz of Colony Collapse Disorder, a dominating narrative in popular culture. This media buzz emphasizes how bees have become increasingly meaningful and visible in public life. We also examine how CCD has framed the urban beekeeping movement in New York City. We introduce and describe two conflicting types of beekeepers that practice "scientific" and "backwards" beekeeping, illuminating the ways that humans conceptualize their relationships and responsibilities toward this insect. In addition, we situate beekeeping within an urban context, specifically, the growing trend toward urban farming and the practices and ideological frameworks connected to greening the city. We look at various ways that city dwellers have positioned themselves closer to some aspects of nature (a murky human concept itself), particularly in light of ecopolitics and a growing interest in sustainability and environmentally green alternatives.

In chapter 4 we examine the intimate buzz where points of contact between humans and honeybees infuse social life. Because we are interested in bees as agentic and active (subjects rather than as nonhuman objects or an "othered" species), we look at how they move humans and affect them emotionally. We then turn to consider how beekeeping is a particularly embodied and emotional experience. Emphasizing the exchanges between beekeeper and bee, we tease out the intimate and sensual nature of tending bees and how humans become connected to them. Because the possibility of getting accidentally stung is so likely, we strip away the sting's various social and personal layers; the sting's meaning to people, and how the fear of stings influences beekeeping practices. We consider how people suit up to work with bees and the levels of protective equipment used—from a full suit, veil, and gloves, to those who literally go unprotected and barefoot. We explore what it means to tend bees without prophylactics, suggesting a more natural and intimate exchange between human and insect (albeit one that has a painful cost). This chapter emphasizes intimate contact zones, specifically the overlap of human and insect bodies. Humans do things to bees, and with them, but what we are most interested in is how bees affect humans through transformations of the self.

Chapter 5 explores the domesticity of bees and how this relates to identity. We examine the ways our informants express and describe the differences between their bees as gendered, and how gender roles are tied to the division of labor. Workers, drones, and queens are narrated by their human keepers as exhibiting typical gendered traits such as promiscuity, beauty, and virility.

Beginning with the swarm, the collective buzzing of racial and ethnic differences of honeybees takes center stage in chapter 6. "Speciation," as explained by Charles Darwin and interpreted by Donna Haraway, is about the distinctions of kind and kin.[45] Animals, including insects, are seen as distinct from one another based on their genealogies, geographical origins, reproductive choices, morphologies, and phenotypic expressions. Categorizing the animal kingdom is mainly about describing variations. Within a family of insect, such as bees, there are many different species. Specific to those species, there are different races. For example, our beekeepers typically kept Italian, Russian, or Carolinian bees, which they described as having specific characteristics or behaviors. The specter of the Africanized honeybee, what has been labeled a "killer bee," looms large in many imaginations. Urban beekeepers, just as humans before them, cast their bees along familiar ethnologic categorization. Through our analysis of these descriptions and larger domestic and international practices that surround honeybees, humans are engaging in national citizen projects.

In chapter 7, we focus on the commercial buzz around bee products, which, unlike pollination, are tangible goods of bees' labor. Honey's sensuality and sweetness feature prominently in folklore and fables and more recently in contemporary holistic health and lifestyle trends. As a substance, it appears to us as untainted, pure, and symbolic of nature's bounty and goodness. Honey is a treat: it is delicious and good for you. While a spoonful of sugar might make the medicine go down, we would scoff at someone who took that advice—but when someone takes a teaspoon of honey, it is probably without comment, and possibly viewed as beneficial. However, honey is a commodity and is part of a nexus of global capital. We look at the ways industry has created multiple revenue-generating innovations for the use of honey and royal jelly—from cosmetics to candles and medical treatments to commercial food production. Honey, beeswax, and venom are among the products humans have used to enhance human "health." Clearly a one-sided exchange where apiary labor benefits human consumption, health, safety, and security, we argue that the work of honeybees is often invisible. Invisible labor, as many feminist scholars can attest, is uncompensated and undervalued and commonly described as "natural" and thus not real work.

In chapter 8, we conclude by moving beyond the bee as an assemblage of connotations, values, and fears, and we advocate for some lifestyle and policy changes that encourage a more just relationship between bees and humans. We examine how human-centered frameworks, and the nature/culture binary, obscures the complex exchanges and interdependencies between species. We have a nuanced relationship with bees—we both admire and abhor them. After all, as the psychologist Hal Herzog has pointed out when considering the complexity of the relationships of humans and animals, "these sorts of contradictions are not anomalies or hypocrisies. Rather they are inevitable. And they show that we are human."[46] As an insect—the idiosyncratic honeybee—lets us fly through and enter conceptual cracks, tiny sites and spaces that open new ways of thinking. Setting aside our humanness, we reflect upon a more ethical relationship to this species and about what forms this could take. It is our hope that *Buzz* affects readers in ways that enlighten and enliven, illuminating the kinship we have to this precious and precarious insect.

3

Saving the Bees

Colony Collapse Disorder and the Greening of the Bee

Where Have All the Bees Gone?

Chances are that you have heard something about the bees' plight and the ominous phenomenon of Colony Collapse Disorder (CCD). Beginning around 2006, CCD has been reported and documented extensively in the American media, reaching wide-ranging audiences and igniting an ecopolitical buzz around the honeybee. The story of CCD has been told in popular yet reader-respected scientific and cultural magazines like *Popular Science* and *National Geographic*, featured in segments on national television network news such as ABC and NBC, and in a wave of documentary films like *Queen of the Sun* (2010), easily accessible online through Netflix. When we mention that we are writing a book about bees, most people ask us, "What is happening to the bees?" "Why are the bees dying?" and "Where did the bees go?" Likewise, Christine, a seasoned NYC metropolitan area beekeeper who sells her extra honey at local farmers' markets, is flooded with these questions by her customers. She said she has gotten so bored and frustrated that she wants to "just make a card to hand out to people so [she] didn't have to repeat the same answer over and over again." People want to know more, and there is more than one CCD story.

The U.S. Environmental Protection Agency describes this phenomenon and "the discovery of a problem" as follows:

> During the winter of 2006–2007, some beekeepers began to report unusually high losses of 30–90 percent of their hives. As many as 50 percent of all affected colonies demonstrated symptoms inconsistent with any known causes of honeybee death: sudden loss of a colony's worker bee population with very few dead bees found near the colony. The queen and brood (young) remained, and the colonies had relatively abundant

> honey and pollen reserves. But hives cannot sustain themselves without worker bees and would eventually die.[1]

Although agricultural records substantiate that bee colonies have gone through periods where they have dwindled and disappeared throughout the last century, many beekeepers and entomologists agree that the current behavior of bees is unique, some would even say alarming, and a few go so far as to say it is potentially catastrophic. To say that CCD has triggered large-scale bee deaths obscures and simplifies what is happening to the bees today. Technically, worker bees disappeared and strangely there was an absence of dead bees, and little evidence was left as to what triggered their desertion of the colony. Because bees simply left their colonies, their bodies were not available to scientists (or anyone for that matter) to conduct postmortem autopsies. Commercial beekeepers didn't encounter a mass of dead bees when they lifted the lids off their hives. They encountered nothing—silence and abandoned combs, and a few lifeless bees. Suddenly, millions of bees were not just sick or dying: they were literally gone. There were clues, but no answers, so the gravity of the problem was undisputed and the mystery compelling.

The media picked up on the inherent mysteriousness of the honeybee's disappearance or, as one *New York Times* article proclaimed, "one of the great murder mysteries of the garden."[2] In this CCD story, the bees didn't get sick and die due to natural reasons (like bacteria) they were in fact "murdered." In the public's eye, this placed the bees at center stage in a media-driven ecocultural detective mystery. The bees eluded the experts, and the rest of us. Did they leave purposefully—simply walk off the job? Were they forced out of their hives?

These important questions were sometimes trivialized and sensationalized, making it seem as if the bees were kidnapped or abducted by some alien force. For example, on July 5, 2010, the television show *Good Morning America* aired a short segment entitled "The Vanishing Bees," detailing how American honeybees have been "disappearing without a trace." A box across the screen declared that ice cream could "go next," if the bees continue their strange behavior. Rather than detail possible causes of CCD and the serious threat to the diversity of our food supply, the focus was on our ice cream supply. However, a cutaway to a Häagen-Dazs ice cream plant assuages concern that ice cream is destined for extinction. Because the company relies on honey to sweeten its boutique brand, it is devoting "millions to bee research." Corporations, like Häagen-Dazs, have a stake in the health and welfare of bees because saving them is a smart PR move. The blurring of ethics and markets is what the posthuman philosopher Donna Haraway describes as lively capital "in which commerce and consciousness, ethics and utility are all in play."[3] The bee's predicament has become a familiar and newsworthy narrative, one that humans have capitalized on.

From the beginning, what was happening to the bees was unclear. Entomologists, beekeepers, and reporters offered different theories as to why bees were disappearing and what this meant for all of us. For a period of time CCD was commonly linked to pesticides, mites, and even cell phone towers and power lines. Regardless of the competing theories and explanations behind the cause of CCD, media narratives commonly drive home the significance of a potential honeybee extinction. One story that gained momentum on the Internet was that Albert Einstein had connected the lives of bees to our existence as a species. Many blogs, such as Global Climate Change, declared that Einstein once said, "If the bee disappears from the surface of the earth, man would have no more than four years to live. No more bees, no more pollination . . . no more men!" Global Climate Change continues, "He wasn't an entomologist, but entomologists around today agree that the sudden and mysterious disappearance of bees from their hives poses serious problems!"[4] The overly simplistic equation, no bees = no humans, paired with the name of one of the most well-known and influential scientists of the 20th century, was quoted widely by environmentalists, journalists, and bee aficionados. An apocalyptical narrative fueled by a dead genius makes for good copy.

Perhaps bees also became such a popular story and a new eco-concern because they have a face. Even though we can't necessarily look into their eyes (like we can with dolphins or pandas), they are more accessible to us as a cause because we do physically encounter them and they are more tangible than a hole in the ozone layer. Bees are ubiquitous; you don't have to travel to a zoo or an aquarium to observe one firsthand. Importantly, we have much to gain from their tireless labor. If bees disappear as a species due to CCD, or some other cause, then our 21st-century American diet is at great risk, as are our global food sources. We could live without bee pollination, but the range of foods that we are accustomed to would be drastically reduced. In other words, there is real cause for concern about bee extinction. The website BeeGuardian.org, citing a U.N. report, concluded that "of the 100 crop species that supply 90 percent of the world's food, bees pollinate more than 70 percent."[5] Apples, strawberries, avocados, carrots, broccoli, olives, onions, and peanuts are on the lengthy list of fruits and vegetables that we commonly eat and have come to expect at grocery stores, all courtesy of bee pollination. As we inspect labels to see if produce is organic or grown in California or Chile, we regard food as being produced by human networks and actions—types of farms and farmers—but rarely by types of bees. We collaborate with bees and employ them but fail to give them credit in a way that takes into account their essential role in our dietary desires and agricultural industries.

This chapter explores the answers to the "what happened to the bees" questions and, importantly, even the questions themselves. That is, why do humans even care? Why is this a news story in the first place? Why is CCD so controversial? We are interested in the opposing ways people approach CCD, particularly the ongoing debates and conflicts within scientific communities and rifts between beekeepers. As reflexive feminist and qualitative sociologists, we examine the construction of scientific knowledge itself and the multiple "contested truths" of CCD. For us, the "truth" is not the point. Our aim is not to empirically "discover" the answer to the mystery that is CCD but to explore the human attention paid to it and how this buzz around bees may be motivated by professional, political, environmental, emotional, or economic interests. While the "Where have the bees gone?" questions are straightforward, it is difficult to provide succinct or satisfying answers to those who ask them. What we are sure of is this: bees are missing. The object of scientific inquiry is gone.

CCD has helped turn bees into very trendy and iconic creatures and they are buzzworthy in the sense that they have become fashionable, a cause célèbre. They are currently more visible in American popular culture and commerce and are welcomed into urban landscapes as part of DIY crafting, local food movements, and rooftop gardening. CCD has also aroused controversy and debate both inside and outside of scientific labs. Bees are now positioned in the realm of animal politics, at the intersections among environmental issues and the politics of food production. The ecopolitical buzz around bees resonates through the contemporary environmental movement, green consumerism, and the idea and practice of sustainability. Many people want to support bees by buying their honey or setting up rooftop and backyard hives or at least communicate that they support them by buying "save the bees, save the world" t-shirts. And surely most humans hope that we can figure out a way to save them—to intervene and interrupt whatever toxic practices we might be perpetrating. Other humans come to the bees' rescue with their practices of interspecies mindfulness. But as with most global environmental crises (e.g., climate change), the scope and interconnected domino effect of the crisis can lead to hysteria. What does it mean if we cannot help the bees and they continue to disappear? If bees do go extinct, perhaps we won't die off as a species, but their absence rattles our human hubris, leaving us more vulnerable in our self-allotted position at the top of the animal kingdom. In their disappearing, bees have become more visible to us.

Defining and Treating CCD

For some people, the name "Colony Collapse Disorder" conjures strong visual images and associations. A local Brooklyn artist and nonbeekeeper we spoke with who was curious about CCD said, "I picture a hive rotting in on itself, like termite wood, like termites got it. The whole thing crumbling or imploding." In his imagination, the "collapse" of CCD was something structural, certainly something breaking down in the bees' environment. That CCD has been labeled a "disorder," rather than, for example, a syndrome, is apropos because a disorder connotes infirmity, either physical or mental. Yet the other meaning of disorder—nonorder, disorderliness, or chaos—also aptly illustrates what is happening to bees as well as the cacophony of human efforts to define, understand, and treat the phenomenon.

From the perspective of bees, CCD can be conceptualized as a type of illness as it is a condition that weakens and threatens their health and survival. From the perspective of scientists working within universities and government agencies, CCD is framed not as an illness but as a disease to be diagnosed and treated. The term "disease" is at the heart of the "medical model"—a theory of intervention that states that the duty of medicine is to discover the etiology of the patient's disorder and then to properly diagnose it. A sick person is unwieldy and unmanageable, but a diseased person can be properly dealt with and issued a prognosis. An illness narrative starts with a sick person's explanation of his or her body's ailments, such as stomach pains and weakness. After a battery of tests that confirm the etiology of the symptoms to be, for example, pancreatic cancer, a sick person becomes diseased or medicalized as a cancer patient. This model is part of a larger process of medicalization, which encompasses the encroachment of biomedical institutions over defining social, personal, and private situations. Medicalized meanings and interpretations—particularly biomedical, Western, allopathic, and genetic—have become the most legitimate and dominant frames in which to understand health and illness. Previously categorized "social problems" such as alcoholism and homosexuality, or predictable life-course events, like menstruation, menopause, and aging, have been open to the medical-industrial complex's expertise. This can be cause for concern as these medical institutions have the power to define our experience for us—they can control the innovation, distribution, and creation of medical treatments. When AIDS went unrecognized or unacknowledged as a serious autoimmune disease, people died.

Diseases are biopolitical designations that are defined, categorized, and ultimately managed by powerful institutions and actors. For humans, these players are made up of doctors and the larger healthcare systems that facilitate our relationships to our own bodies and sense of personal well-being. Entomologists and biologists, working for universities, industrial farms, and pesticide companies, have battled over the causes and the cure for CCD as a disease. Perhaps, not surprisingly, CCD has been referred to as a form of "insect AIDS" in the media; an example is a *New Yorker* article that reported how some researchers had "concluded that the bees' immune systems had collapsed. It was as if an insect version of AIDS were sweeping through the hives."[6]

An article on CCD published in the prestigious medical journal *The Lancet* proposed that bees "could be more susceptible to disease when their immune systems are weakened by antibiotics or stresses caused by apiary overcrowding, poor nutrition, or migratory stress, because commercial bees are often transported over long distances for pollination."[7]

The notion that bees have immune systems, like humans, points to the ways in which CCD is assumed to be a disease. As the anthropologist Emily Martin describes in her book *Flexible Bodies* about AIDS and immunity, in the 1990s immunity metaphors circulated in medical and popular journals, and they signaled a shift in how we conceptualize our own relationships with our immune systems.[8] AIDS was increasingly described by medical specialists and visualized in the media as an internal battle between the virus and our bodies—in essence promoting the idea that HIV-infected bodies were at war with something inside them. Likening the bee's immune system to our immune system in itself is significant because it illuminates how bees are situated within human cultural, scientific, and medical frameworks that theoretically have the power to offer definitive explanations. If CCD is likened to an autoimmune disease then scientific institutions are potentially capable and ethically compelled to find a cure. Bees, like humans, become subject to the medical model.

Articles speculating the causes of CCD circulated for about five years until it was reported that biopolitical experts had likely found the source of the problem. A *New York Times* story that ran on October 6, 2010, revealed that entomologists and Army researchers had "proven" that a "fungus tag-teaming with a virus" had led to massive bee deaths.[9] A group of scientists led by Jerry Bromenshenk of the University of Montana in Missoula and the Army's Edgewood Chemical Biological Center outside of Baltimore joined forces to investigate the problem of CCD. According to the *Times* article, the Montana/Baltimore team used analytic software employed by the Army. The software is described as "an advance itself in the growing field of protein research, or proteomics . . . designed to test and identify biological agents in circumstances where commanders might have no idea what sort of threat they face. The system searches out the unique proteins in a sample, then identifies a virus or other microscopic life form based on the proteins it is known to contain."[10] Charles H. Wick, an Army microbiologist, is quoted as saying that the mission of the biological testing center is "to have detection capability to protect the people in the field from anything biological." Significantly, CCD fell under the same rubric as toxic and lethal biological agents such as anthrax, bacteria spread by herbivores that could be used as biological weapons.

Again we see that viruses and fungi are empirically identifiable, verifiable, and, ostensibly, manageable. The answer for the government is to protect bee colonies by focusing on the fungus, which is "controllable with antifungal agents."[11] In other words, they argue that just as human beings are susceptible to viruses and fungi, bees are as well. Of course, not all researchers or beekeepers accept the Army-sponsored study's conclusions about the likely cause of CCD. In fact, the first commercial beekeepers who observed CCD made connections between the location of their hives to fields treated with new insecticides such as neonicotinoid imidacloprid. The beekeepers claimed that CCD affected colonies a few months after the first neonicotinyl exposure. They surmised that bees were bringing the insecticide back to the colony through contaminated pollen and nectar, which was then building up over time. However, to date "U.S. regulators have dismissed beekeepers' on-the-ground evidence" that neonicotinoids are connected to CCD.[12] According to the environmental sociologists Sainath Suryanarayanan and Daniel Lee Kleinman, "Government officials view beekeeper evidence as anecdotal, and they will not consider it in promulgating regulations, since beekeepers do not isolate causal variables in the way done in formal laboratory and field experiments."[13] The beekeepers' insecticide hypothesis has yet to be scientifically proven.

Other researchers have suggested that while both insecticides and the fungus-virus combo may make bees vulnerable, it is still probable that the environmental factors of habitat erosion, stress, and pollution also play a role. In his book *The Backyard Beekeeper*, Kim Flottum explains that CCD stems from a combination of pesticides, nutritional distress, climatic aberrations, *Varroa* mites, and nosema, which has weakened bees and made them vulnerable. Flottum asserts that CCD lasted for four years and is now mostly over, thanks to prevention methods such as "maintaining good bee nutrition all year long" and "avoiding monocrop sources of food and agricultural pesticides."[14] His perspective on CCD is holistic, environmentally situated, and, we would argue, more complicated and messier than the fungus hypothesis.

Flottum's explanation of CCD as the result of a host of environmental stressors is, however comparable, once again to a *human* condition known as environmental illness, or multiple chemical sensitivity (MCS). This illness is expressed in an often elusive and complex array of symptoms and disorders. For example, one person may exhibit memory loss and a facial rash while another could have severe headaches, shortness of breath, and nosebleeds. A person with MCS cannot be placed within the medical model. As the sociologists Steve Kroll-Smith and H. Hugh Floyd observe in their book *Bodies in Protest,* MCS has not been recognized as a disease—it escapes medicalization because of the plethora of triggers and symptoms it encompasses.[15] Some people (both sympathetic doctors and actual sufferers) believe MCS is caused by human intervention into environments and the "chemical revolution." This revolution is the result of modern society, including the building materials used for malls, universities, and houses, as well as consumer goods, such as perfumes, household cleaning products, and deodorants, that make certain people sick, while others remain inexplicably immune. MCS affects those who suffer from it in unexplainably different ways and is triggered by a variety of unrelated chemicals at doses below levels deemed to be safe for the general public. From a medical perspective, MCS is difficult to make sense of. However, given the increase in the types of synthetic and chemical compounds we come into contact with in everyday life, from antibacterial soap to plastic covers meant to protect cellular phones, it is plausible that some bodies could become vulnerable to the complicated artificial environment that surrounds us.

MCS is an elusive human syndrome that occurs under stressful environmental conditions. Similarly, CCD is an elusive insect syndrome. As an interspecies comparison, CCD may be akin to MCS for bees, whereby bees become weakened by the demands of their lifestyle (particularly monocrop pollination) and the toxic load that is in the environment, so that their ability to resist mites and viruses is weakened. Complex conditions like these highlight the postmodern demands of fast-paced, technologically driven global economic systems and the ways in which some living organisms (including bees and humans) are unable to adapt to them. This is just one of many human theories that attempts to place CCD within a larger ecological system. While CCD may be almost solved for some, others still contest and debate the mystery of the bees' disappearance. In fact, the plight of the bees has even reached the macabre.

Another explanation of the cause of CCD was widely reported in early January 2012: arguably, it is framed as literally and metaphorically horrific. As declared in a *Scientific American* article, a "zombie fly parasite" is killing honeybees.[16] The discovery of the parasitic fly was somewhat accidental, occurring when a San Francisco State biologist named John Hafernik left some bees in a vial in his desk "and forgot about them" (they were intended as a snack for a praying mantis). When he went to check the vial he was "shocked" to see fly pupae surrounding the bees. According to an Associated Press report, "the fly deposits its eggs into the bee's abdomen, causing the insect to walk around in circles with no apparent sense of direction."[17] This confused state is what turns bees into "zombies." But the story gets even more compelling, like a bee version of *Night of the Living Dead*. After the bee is invaded by its parasitic host, it stumbles around confused and then dies, "as many as 13 fly larvae crawl out of the bee's neck . . . similar to that of ants that are parasitized and then decapitated from within."[18] As noted by many scholars of popular culture, zombies, the walking ghoulish undead, have enjoyed a "renaissance" in the early 21st century by appearing in a range of movies, TV shows, and novels.[19] In this frame, reports of parasitic flies and bee zombies locate CCD in a landscape of nature, science, and popular horror tropes. This CCD narrative appeals to horror fans, but the scientific jury is still out because the fly could be another stressor pushing bees over the "tipping point" or "it could play a primary role in causing the disease."[20]

If the parasitic fly hypothesis is validated through further research, some scientists may advocate for isolating hives and killing the so-called zombie fly (protecting bees from tiny terrifying insect invaders). Likewise, eradicating viral and fungal agents clearly delimits CCD as a disease and positions it under a microscope, a location of scientific control. Those who conceptualize CCD as part of a greater environmental crisis, one that threatens humans and implicates us as part of the problem or solution, have to wrestle with multiple locations, sites, and stressors that resist tidy fixable categories. Fighting a fungus allows for a fixed target and an obvious rational solution (fungicides) as does honing in on a zombie fly; fighting global pollution, soil toxicity, and urbanization is much more difficult. These are unquestionably human problems, infinitely difficult to disentangle.

Because there is so much at stake financially (i.e., the food supply chain), there are reports of human intrigue and conspiracy in the study of CCD. In the words of one scientifically minded beekeeper, "The infighting and competition for funding, the backstabbing, the bogus tricks to grab attention and credit are part of what prompted me to stop writing about science and bees." He told us that he "could not stomach the intrigue any longer." One of the most well-publicized stories of conspiracy centered on Jerry Bromenshenk, who spearheaded the fungus hypothesis and the military use of bees (further explored in chapter 7). Bromenshenk was featured in the documentary *Colony* and has been widely interviewed by the media, appearing as an expert guest on CBS television and radio. Previously, he received funding from Bayer Crop Science (a subsidiary of Bayer pharmaceuticals and a major pesticide corporation) to work on a CCD project. However, Bayer Crop Science was subsequently accused of poisoning bees and was involved in a class-action lawsuit brought about by thirteen North Dakota beekeepers.[21] According to an online CNN report, "He had signed on to serve as an expert witness for beekeepers who brought a class-action lawsuit against Bayer in 2003. He then dropped out and received the grant."[22] The article argues that Bromenshenk's company, Bee Alert Technology, "will profit more from a finding that disease, and not pesticides, is harming bees."[23] The Bayer/Bromenshenk court case can be understood in the words of one of our informants who asserts that "all the theories and unusual claims prove that the correct name for CCD is not Colony Collapse Disorder, but Cash Cow Discovery." While this story implicates conspiracy theory, it illustrates how bees are arguably the most politically valuable and contentious insects of the 21st century. A wide range of people with disparate interests, from local farmers to Army scientists, are the stakeholders in bee futures.

For the time being, the mystery of CCD continues to reverberate and buzz outside of news media outlets. Because humans are implicated in its cause and extinction narratives are titillating and frightening, CCD has become a dominant theme in documentary films as well. Aside from the bees, beekeepers themselves are sometimes shown as curious and eccentric. An informant lamented about the oddball framing of his colleagues in the film *Queen of the Sun: What Are the Bees Telling Us?* (2011), noting that "the general impression it gives is that beekeepers are a quirky bunch, as the filmmaker sought out characters that would be at home in a Werner Herzog flick." He added that they "were oblivious to how silly they seemed, or were long past caring." Other recent films distributed for wide release include *To Be or Not to Bee* (2010), *Colony* (2009), *The Vanishing of the Bees* (2009), and *The Silence of the Bees* (which originally aired on PBS in 2007). The latest spate of documentaries on bees demonstrates that a growing section of the lay public is truly curious about what is happening; these films have multiple audiences beyond bee aficionados and environmentalists. Although the lives of bees are fascinating enough on their own, CCD narratives helped bees achieve a level of popularity that is unusual for insects. For now, bees are no longer relegated to nature specials and NOVA documentaries. In the words of one beekeeper we interviewed in the prosperous and recently gentrified brownstone neighborhood of Park Slope, Brooklyn, "bees are the new cupcakes," referring to the trendy cupcake cafés and bakeries that have popped up in the neighborhood. Perhaps it is because the bees are disappearing that, as an insect to fetishize, they seem to be ubiquitous.

Something happened to honeybees that was empirically verifiable (i.e., missing workers and abandoned hives). However, even though some veteran beekeepers have convincingly argued that CCD should not be identified as a singularly and historically pernicious disease because bees have always battled a variety of diseases, our goal here is not to take sides and weigh the merits of empirical or anecdotal evidence to bring you the definitive explanation of what CCD is. Because this is an emerging syndrome, humans are still trying to precisely explain what is happening. The meaning of CCD is still in play and negotiable. People often talk about it in ways that are discernible to humans (i.e., parasites) and how we narrate the disease is significant. The way we describe CCD says something about *us* because we have a stake in bees. Some species go extinct and we may never hear about them, but the bee problem is indeed our problem—it is more than a story for the headlines.

"It's All about the People, Not the Bees"

In trying to better understand CCD, we think it would be useful to look more closely at the beekeepers themselves. While there are probably several different types of beekeepers in New York City, in our research two schools of thought on beekeeping surfaced: what we initially referred to as the rational/scientific and the naturalist/backwards paradigms. We have come to use the more simplified categories of "scientific" and "backwards" for purposes of brevity and as a way of framing beekeeping practices that are based on similar beliefs. The term "backwards" was coined by beekeepers themselves, and while it seems pejorative, it refers to going back in time to a preindustrialized era before commercial beekeeping. The backwards practice and philosophy encourages a more hands-off approach, literally to leave bees alone and accept that some will die off. Dead bees are bees "being themselves," as many a backwards beekeeper will note, for better or worse, and ultimately dead bees are preferred as compared with over-intervening in their lives (particularly through the use of chemical agents). "Backwards beekeeping" was first coined in 2001 in a *Bee Culture* magazine article and later used as the moniker for the L.A.- and NYC-based Backwards Beekeepers chapters.

In comparison, scientific beekeepers are more invested in helping bees through empirical observation, intervention, and technology. Scientific beekeepers tend to be well trained, use traditional human-made beehives, and practice beekeeping according to commonly accepted procedures that have been in place since the 19th century. These beekeepers tend to emphasize the need for human intervention, use treatments if necessary, and advocate a judicious hands-on approach.

Of course, beekeepers do not all conform and fit into these designations, as there is a noticeable blur with regard to certain issues and treatments. They are conceptually helpful to us, but, importantly, the beekeepers we spoke with sometimes elude them and creatively cut and paste their beekeeping styles. For example, Tim O'Neil, a Brooklyn beekeeper, instructor, and blogger in his late twenties, told us that he was a centrist of sorts: "I am a *center left beekeeper* which means I try to avoid treatments when I can, but I use them when I think I need to. But I try to use nonpharmaceutical treatments." O'Neil, as well as other beekeepers, cull from both traditional and more radical beekeeping practices. However, there is observable divisiveness in the NYC beekeeping community. The main issues that divide scientific and backwards beekeepers are about what the best way to keep bees is and, relatedly, what the root cause of CCD is. As will become apparent, the questions of *practice* and *cause* are interconnected. The explanations and solutions behind each of these individual concerns will lead you back to the other.

The groups, while not actually interacting with one another in real time, are in dialogue with one another. For example, those within the scientific community often ridicule the backwards beekeeping practice of "communing" with the bees, which some consider trendy and silly. As we discuss in chapter 4, communing with bees can take the form of listening to bees, watching them, and sometimes ascribing personalities that may indicate emotional attachments. It also suggests what some have referred to as the "magic" of the hives and the experience of being with bees as meditative—a medium of transcendence and awe.

As with other cultural practices, from tattooing to mushroom collecting, the Internet changed beekeeping culture. According to Jim Fischer, the NYC Beekeeping codirector, when the Internet beekeeping network was formed in 1995, "*experts* with two hives and six months of experience" could trade and relay information. Skeptically, Fischer believes that "like-minded people found each other" and "convinced themselves that they are right, tossing aside the science." Since the explosion of beekeeping, various groups have formed that delineate themselves with specific language, ideas, and management programs.

Among the backwards beekeepers there are different types of practices, such as organic beekeeping, biodynamic beekeeping, and holistic beekeeping—each founded on related but distinct ideologies. According to Fischer, who is aligned with the scientific group, these approaches are equivalent to "cults" and "all have different dogmas." Fischer continued: "It's not enough that they keep bees, they have to do something special," hinting at how backwards beekeepers integrated bees into their lifestyles as some form of trendy green fashion statement. He stressed that beekeeping "did not need a fancy word in front of it": it was just "beekeeping." His concern with these "dogmatic bee cults" mainly stems from the fact that they are, in his view, not practicing integrated pest management. Keeping bees pest-free and healthy is serious business to some beekeepers who do not welcome alternative practices.

The scientific beekeeper is characterized by members of the NYC Beekeeping Meetup group. This group's beekeeping approach is based on empirical observation of bees themselves, closely connected to the work of entomologists, and is dedicated to educating the public free of charge. This more conventional community is on a mission; it is highly professional, organized, and utterly committed to spreading information about bees and beekeeping. They use scientific language to talk on behalf of the bees' well-being and see this as "public service" for sustaining bees in NYC. In our fieldwork, we witnessed a beekeeping ethic among members from this group. There is clearly a right way to keep bees and this was a mission for those of us called upon to become citizen scientists. It was generally assumed and understood that for most of the members beekeeping was indeed a hobby, as Fischer stated jocularly that we all should "breathe" because "we are not here to make a living" and "we are supposed to be enjoying it." Yet many of the lectures conducted by the meetup group referred to studies by entomologists and often included PowerPoint presentations that relied as much on scientific graphs and charts as on personal photographs of bees. We were encouraged to closely watch and document bee behaviors, such as the absence or presence of flight activity in a hive, and to record our observations as data. The committed scientific beekeeper doesn't simply hang out with the bees and have a beer: she carefully observes, monitors, and records what they do. In theory, this seems to suggest that if something goes wrong in the hive, it can ostensibly be remedied by making empirically based decisions and adjustments.

The backwards beekeeping community is not as organized as the meetup group and is composed of mainly younger people (those in their twenties and thirties), with a more overtly political message or attraction to creative communal living. In some cases, bees become other housemates, living among a tribe of what a recent college grad described as "artists, vegetarians, queers, cyclists, and students" united under one roof. Often members of the naturalist beekeeping group are simply interested in bees as a direct way to connect with nature within an urban environment. Some naturalist beekeepers see bees as part of other overlapping interests and values, such as veganism, holistic health treatments, rooftop and backyard farming, and other green lifestyle and political choices.

We visited a collective in Brooklyn that half-mockingly and half-seriously identified itself as "the federation." The federation's home is a four-story brownstone in the predominantly black neighborhood of Crown Heights. Seventeen people shared the space; however, only a few people were responsible for the bees. The hallways were filled with bikes, and the common area had DJ equipment and a carefully curated, likely ironic, black velvet unicorn painting on the wall. Sipping on what was described as "bee liquor" (a form of mead) brought back from a European tour from a friend who is a professional musician and enthusiastic new beekeeper, we spoke with her and four federation roommates (all in their twenties) about how the bees fit in with their lives. Everyone agreed that beyond bees, they were all deeply interested in reducing their ecological footprint, creating a more sustainable environment, and eating locally cultivated food. They did not use chemicals to treat for mites and did not talk about bees in scientific ways. This is a more collective and DIY effort that speaks to how bees are welcomed into alternative and green communities.

The backwards approach is epitomized by the beekeeper Sam Comfort, who is the founder of Anarchist Apiaries. He operates on the outskirts of metropolitan New York in the Hudson Valley and has trained many urban beekeepers. Comfort, who is on a mission to "save the bees" and "save the state," travels throughout the five boroughs for beekeeping events. In figure 3.1, he is entertaining a crowd while discussing top bar hives and natural beekeeping practices at the Brooklyn Botanic Garden Bee Day Celebration in 2010 at Prospect Park.

In comparison with the NYC Beekeeping Meetup group, in a presentation by Comfort we were introduced to the practice of beekeeping as part of a "natural anticorporate approach" to "work to get a more natural wild world." While Anarchist Apiaries is outside the city, Comfort has teamed up with an L.A. beekeeper named Kirk Anderson, as well as Brooklyn-based Meg Paska to form a chapter of backwards beekeepers in New York City. Anderson borrowed the idea of backwards beekeeping from an article in *Bee Culture* magazine written in 2001 by Charles Martin Simon. Entitled "Principles of Beekeeping Backwards," Simon writes, "I have established mystic contact with the spiritual core of apiculture, and now anything is possible. Some of you old timers might resonate with this statement, but most of you, I'm sure, will not have a clue. Many will be irritated by what you perceive to be my arrogance; but, you have it backwards."[24] In the piece he is critical of tried-and-true beekeeping practices, from the use of a Langstroth frame (the traditional square box hive) to treating bees for diseases. After forty years of

Figure 3.1. Sam Comfort singing at the Bee Day Celebration. (Photo credit: Mary Kosut)

struggling with bees and playing by the rules, Simon advocates to "leave your bees alone." He argues that dead bees are a part of life: "I have observed that the harder you fight to keep your bees alive, the faster they die. Cut them loose, give them freedom, the freedom to die as well as the freedom to live, and they live better."[25]

At the first meeting of the NYC Backwards Beekeepers in July 2011, Comfort explained his perspective on CCD. He argued that there is no such thing per se, shifting the focus from the bees back to the humans, those responsible for naming CCD. Like others, he put humans at the center of CCD, advancing the idea that its emergence and labeling say more about human behavior than insect behavior. Comfort passionately asserts, "It's PCD—People Collapse Disorder—bees are just a vehicle. . . . It's all about the people, not the bees. It's a rift between us and the ecosystem and monoculture. Bees are pissed off. . . . They want different kinds of pollen." His argument is based on his experience spending years working for professional agricultural beekeepers who pollinated crops mainly in the western United States. Two crops that depend on trucked-in bee labor particularly are almonds in California and cherries in Washington State. He observed that bees didn't have much else to pollinate beyond the crops themselves—no wildflowers and plants. This means that their "diets" were the equivalent of a human being eating only one or two types of the same food with no other nutritional sources—you can stay alive but you won't necessarily thrive. He explained his philosophy: "There are two things in life, monotony and diversity—bees need diversity. They are foragers. They strive on chaos, which is basically just a whole lot of stuff going on. They have been domesticated for 150 years but they are not like cows. They haven't been messed with in the same way." For example, dairy cows have been modified through the injection of hormones, which increase milk production. Still, the bees have been "messed with," particularly with monocrop pollination. Monocropping is an agricultural system in which the same crop is grown over acres of land year after year without rotation. This is very efficient but causes the soil to lose nutrients, leading to the use of more fertilizers and pesticides. The ground and the crops produced are chemically saturated and affect the local ecosystem. Monocrop production poses many potential risks for species directly affected, like humans and bees.

The scientific approach as explained by Jim Fischer is an interesting comparison. Like naturalists such as Comfort, he believes that there is a people problem too. Fischer strongly asserts that "the claims about cell phones, power lines, GMOs, and so on are simply craven bunk. These are people with pre-existing axes to grind, who are cynically exploiting the reports of the problems of bees and beekeepers to further their own agenda, and, more importantly, to make bees not just the mascot for the environmental movement but the current 'charismatic megafauna.' " He continues: "To really understand the problems of bees and beekeepers, it helps to step back and look at bees and beekeepers as one small part of agriculture as a whole." Fischer passionately asserts that the problems come down to "modern times" and "invasive species of pests and diseases as a result of world trade":

> What people call Colony Collapse Disorder itself is nothing more than a combination of two or three diseases of Asian bees that made their ways to our shores in uninspected and unregulated world trade. Invasive species are plaguing all of agriculture everywhere, as it seems that the combination of containerized freight and fast ocean crossings by modern

> cargo ships are allowing pests and diseases to spread from warmer places in Asia where low-cost "consumer goods" are made, to everywhere else. . . . Then containerized multi-mode freight became common, with containers that could be handled by giant cranes to unload entire ships in hours, and placed on train cars or truck flatbeds to be deliverable in days rather than weeks. This allowed swarms of bees to fly into a container in Indonesia, and emerge alive, but carrying an invasive disease, in Idaho.

Like the naturalist beekeepers, CCD is situated within a particular historical and ecological moment; however, the issue of mites and diseases is posited as the most pernicious and threatening force. So here the danger to bees is not just humans, as it were, but a destructive relationship between pests and humans that has formed because of globalization and commerce.

As Fischer explains, from his standpoint CCD, while threatening, has been fabricated as too complex and blown out of proportion. Experienced beekeepers like Fischer have seen other syndromes harm bees. Even though it was not as widely publicized as CCD, foulbrood was seen by beekeepers as a serious threat to their colonies back in the 1980s. Foulbrood is a contagious bacterial infection that affects bee brood, a term that describes three early stages of bee growth—egg, larva, and pupa. Foulbrood is particularly harmful in the early larval stage of development. Adult bees can carry foulbrood spores without becoming sick, but those bees can wipe out an entire colony by unwittingly transmitting the disease to vulnerable larvae. Foulbrood still exists, but it is treatable via antibiotics such as tylosin tartrate (Tylan Soluble) and oxytetracycline hydrochloride (Terramycin) that can be mixed with powdered sugar and dusted on bees so that they ingest the antibiotics.[26] The larger history of bees is a history of struggle for survival, particularly against mites and viruses. From this standpoint, the cause of recent colony deaths can be linked to two constants in the life of bees: modernity and disease. However, there are other explanations.

The treatment of mites, through chemicals or natural substances, is another issue that divides beekeepers. Mites are very small anthropods, invertebrate animals with external skeletons that are members of the taxa arthropoda, including insects, arachnids, and crustaceans. One of the most diverse and understudied populations of organisms, parasitic mites—those that feed off the host organism at its expense—can harm agriculture, rodents, birds, and humans in the form of chiggers and scabies.[27] For humans, parasitic mites such as these are insufferable, the microscopic critters can cause excruciatingly itchy and unsightly red bumps, and they must be medically treated. Yet some humans believe that bees can and should wrestle with their mite problems themselves, without our intervention. Backwards beekeepers, like Kirk Anderson, advocate a hands-off approach to mite control, as "the biggest pests that bees have is man, not mites. Bees are pretty smart—so we let them make their own decisions. Let the bees manage the mites." For some self-described backwards beekeepers, mites and bees should coexist and the bees themselves have the ability to stave off the mites. The bees are empowered to naturally respond to mites, even though that may mean that an entire hive will die as a result of the lack of human intervention. In comparison, the following quote from a scientific beekeeper illustrates how some people see an inherent ethical responsibility in tending and monitoring bees:

> By trying to be as detached as possible, the beekeeper, like a doctor, can do a better job of not misinterpreting the behavior of the bees. As the decisions of the beekeeper can have a profound impact on the future of the hive, the beekeeper owes his or her bees a dispassionate and professional evaluation, and the best that science has to offer in the way of help when they need it.

For some, scientists know what's best for hive health, and the potential cure for any disease can be found within modern scientific developments.

The differing theories of what CCD is—where it originates, why it exists—is the scaffolding on which beekeepers build their everyday practices. Choosing to treat for mites (with powdered sugar or chemical agents) or choosing *not* to treat for mites becomes a major dividing line for some beekeepers. What this points to is debates over levels of intervention and how much intervening the bees presumably need by people in order to continue to thrive. Both camps agree that humans are the primary problem, but they are divided when it comes to what role they should play in fixing it. This is in part because many naturalist beekeepers see bees as wild animals—beings that should be free of domestication and live naturally out of the realm of human control. This version of wild is what the anthropologists Sarah Franklin and Margaret Lock describe as "old wild," a landscape where feral animals like wild geese and boars roamed free, rather than the "new wild," a place populated by creatures that have emerged amid regimes of biological control and bioengineered life forms such as Dolly (the famous cloned sheep) and the crossing of genetically modified plants with nongenetically modified ones.[28] For some self-labeled anarchist or backwards beekeepers, wild and feral as in the *old wild* bees may be the answer to CCD. Bees are ideally not bought and shipped in boxes from Georgia or Florida, which is how some beekeepers in the New York City area acquire them. Instead, they are "rescued" when they naturally swarm in the area, resulting in bees that are familiar with local conditions—climate, ecology, water, and food sources. Comfort explains why a natural swarm is best for the bees:

> A natural swarm is like if you and a bunch of friends decide that you want to hang out and live together, live collectively doing the things you like. Bought bees with a queen that come in a box are like you and a bunch of people who are stranded together on a bus and you have to figure out how you are going to get along. Bees can get along on their own—this is what they should do, when we intervene it's like we are forcing a bunch of bees to get together. It's just off.

Kirk Anderson endorses Comfort's bus swarm idea, arguing that bees should swarm on their own and not be forced into living together. He also asserts that "people should let bees die, no chemicals, no interventions." Anderson only feeds bees once when they are in a new home (a half-and-half mixture of cane syrup and water) and then "lets them alone." He is on a mission to make beekeepers use feral or wild bees.

The idea of being "green" or even the notion of nature itself is a social construction, as are the concepts of "wild" and "feral." A wild, untamed nature has historically been seen as a form of nature that is out of control. As the sociologist Matthew Immergut argues, there is a dark side to feral nature that we must tame in order to keep it at bay.[29] He compares the practice of manicuring lawns and picking weeds to a recent trend of male hair removal, deemed manscaping in the media. A perfect weed-free frontyard is akin to a smooth torso and back, sculpted, ordered, and enticing; an overgrown lawn, like overgrown chest hair, becomes a signifier of laziness and failure. Nature in these instances is grotesque; it makes us feel uncomfortable and even unclean and uncivilized. Conversely, it is asserted that beekeeping has been made to be too civilized and too ordered. When bees live without a man-made box (known as a Langstroth), they form rounded and misshapen hives—they don't innately design their existence within a square or rectangle. While the combs are "mathematically" made and organized, bees are not linear builders. They are adaptable architects and coinhabitants. They will live in mud, and they thrive in abandoned foreclosed homes in Florida. We have heard stories of bees surviving icy winters in upstate New York and living in snowdrifts in Newfoundland.

The notion of rescuing bees contradicts the ideal that they are feral to some extent. If they are truly feral, then they should be able to survive without the well-intentioned actions of humans. However, it's obviously not that simple because the relationship between humans and bees has never been uncomplicated and has never been equal. At this point in history, we find ourselves asking whether bees actually need us in order to survive as a species. According to Jim Fischer, the answer is unequivocally yes: "Some people say that beekeepers are bad people because bees are wild animals. But if not for beekeepers, then all bees would be dead. And beekeepers are thought to be bad because bees should be wild and they shouldn't be captured by humans. But bees would not exist without human help. Beekeepers are necessary."

Fischer's perspective, as well as that of many others who fall under the scientific beekeeper paradigm whom we have talked to, represents a particular ecological position in which humans see themselves as the stewards of the bees. Their role is to help nature, and in essence they are caring scientists who are out to protect and preserve the bees. This stance is aligned with some strains of the Deep Ecology movement that emerged in the 1960s as part of the larger environmental movement. Deep Ecology advocates "ecocentric rather than anthropocentric" values based on the idea that "wild ecosystems and species on earth have intrinsic value and the right to exist and flourish."[30] Importantly, many of these beekeepers do not see their actions as "controlling nature" per se, but assisting the natural world that has been threatened by human domination. In contrast, naturalist beekeepers have a more radical ecophilosophical approach that may also be rooted in some tenants of Deep Ecology, particularly the idea that "wild ecosystems" should be not only protected but also *restored*. For people like Comfort and Anderson, self-described "backwards beekeepers," the restoration of bees to their previous "wild" state is key to their survival. This means radical removal of human intervention ("let the bees be bees"), so that the insects ostensibly become more "animal-like."

While we are not qualified to argue what is best for bees, more protection/intervention or a hands-off approach, it seems clear that the dominant systems of agricultural production, including monocropping, are not healthy for bees. Creating a more naturally bee-friendly environment through organic or alternative farming practices might benefit bees the most, but ultimately this type of intervention would require radically changing our relationship to food and the planet. Ceasing to use chemicals, both fertilizers and pesticides, advocating local smaller-scale food production rather than corporate farms, and promoting biodiversity would mean that humans would have to go "backwards"—back to the land before there were genetically modified crops and fungicides. For environmentalists, this idea is the most ethical and sustainable way to produce food. For others, going backwards is tantamount to advocating for communism in 1950s America, or living on a hippie kibbutz; both essentially are ideas that are out of touch with modern life. If we went backwards, the shelves at Wal-Mart would not have as many enormous fruit and vegetable pyramid displays and prices would be higher—there would be less for more, rather than more for less. But our affordable, abundant, and diverse food supply comes at a high cost if you factor in the harm done to other species, especially bees.

Greening the City

As reports of CCD began to circulate widely in the media, another trend occurred in major American cities such as L.A., San Francisco, and New York. Urbanites began keeping bees within cities themselves, rather than trekking to the outskirts or well-isolated rural spots.

Mainstream local and national presses quickly picked up on this trend. There are many elements of "human interest" to fuel coverage of the beekeeping phenomenon aside from the saturation of CCD stories and potential bee extinction. Within the context of a green-friendly cultural climate, keeping bees in the city is an act of cultivating and bringing "nature" (assisting pollination) into environments that ostensibly need it. According to the writer of one *New York Times* op-ed, local urban honey is good not just for allergies but for "the health of the city" as well: "Take the honeybees of East New York Farms! an organization of urban farmers and neighborhood farmers' markets. These Brooklyn bees pollinate crops for the entire neighborhood. They aren't just making honey: they're building community, creating income and employment and maintaining vital urban green space."[31]

There is also an inherent quirkiness to housing bees on rooftops and concrete patios, because they seemingly don't belong there. As shown in figure 3.2, a group of young beekeepers in Crown Heights, Brooklyn, silently inspects hives as an elevated subway train clanks by about a hundred yards in the distance. In the background the Manhattan skyline peeks through a layer of midsummer haze and smog. With approximately nine million human inhabitants fighting over space and resources, somehow the bees manage too. While some green treetops are in sight (as possible pollination sources), there is no discernible natural water source anywhere close. Bees need water and the East River is miles away, so they adapt to alternatives like fountains and kiddie pools.

The urban landscape has many locations in which honeybees can become even more meaningful and visible to people. The growth and popularity of urban beekeeping may tell us just as much about human problems and desires, as it does about bee problems, their current plight, and the threat of CCD. Even though bees have long been New York City residents, people are cultivating relationships with bees and inviting them into unlikely hyper-urban settings. Bees as semiferal and autonomous insects signify "real" nature, and the possibility of their survival in grimy, congested, and toxic neighborhoods is a sign that humans are doing something right and generous. In our fieldwork we found people tending bees in unlikely spaces: on rooftops of synagogues; in backyards in bustling Brooklyn neighborhoods; on the tops of MTA and Port Authority truck transport stations; and in Bronx

Figure 3.2. Beekeepers on a Crown Heights rooftop with a train in the background. (Photo credit: Lisa Jean Moore)

community "farms" amid the diesel smells from Hunts Point. Figure 3.3 shows beehives on Randall's Island, directly next to the Triboro Bridge, also known as the RFK Bridge (in the background) which moves tens of thousands of vehicles daily from Queens and Brooklyn to the Bronx and back again.

Cities call to mind clogged traffic, scaffolding, and sidewalks, rather than beehives and chicken coops. We found that people who kept bees in New York City were often genuinely interested in housing chickens as well. Two of our Brooklyn informants started with beehives, then moved onto chickens, which they house in urban backyards on slender plots of land bounded by concrete patios, wire and wood fences, and neighbors in every direction. One person even joked about bees being like a "gateway drug to chickens." Pushing the idea of urban farming even further, another informant who has bees, chickens, two dogs, a cat, and a frog said she also has considered getting a goat but they are not permitted under city ordinances.

Notwithstanding, urban farming has its limits; most neighbors approve of gardens and will tolerate bees but animals such as goats, pigs, and cows have yet to be welcomed into the inner city. The reluctance is likely due in part to bans that arose after World War II, as "cities, seeking to eradicate any traces of agriculture within their limits in order to show they were full-fledged municipalities, forbade the raising of livestock, chicken and other creatures used in food production."[32] Certain animals, those seen as potential food, like chickens, breach the boundaries that divide urban from rural. When plucked out of a rural context, they can appear as a nuisance, threat, or anomaly. We welcome only

Figure 3.3. Beehives on Randall's Island with the RFK Bridge in background. (Photo credit: Lisa Jean Moore)

certain nonhuman species to share our sidewalks and other collective city spaces, especially in the case of contiguous backyards.

While most neighbors might view beekeeping positively in the abstract because there is a human and environmental benefit, and the presence of a lone single bee careening about a backyard is typically benign, a hive can be a bit suspicious if not threatening. One or two bees appear in the realm of human control, but hundreds or thousands of bees, as a group or colony, become more of a natural force that someone could not wrangle physically or emotionally. Hives may illicit anxiety because, as previously discussed, bees can and do swarm and can and do sting. And for the uninitiated, a flying mass of a thousand bees can be an absolutely frightful experience. It is terrifying mainly because we don't know what they are doing, and we rarely come face to face with large numbers of mobile bees, uncontained en masse. Regardless, before beekeeping was legalized in NYC in 2010, there were numerous editorials and articles in local papers written in favor of making beekeeping a legitimate practice. The potential harm was downplayed and narratives of happy well-managed bees dominated media accounts as illustrated in this excerpt from a *New York Times* piece about a father and his nine-year-old son who kept bees in the City Island section of the Bronx:

> In a demonstration of beekeeping practice, Mr. Gannon and his son filled a hive with smoke, using a bee smoker, a small device equipped with a hand pump. Smoke calms the bees, but also makes them anticipate having to abandon the hive because of fire. They gorge on honey, in preparation for a quick exit, and, like humans, they mellow out after their big meal. It makes them less likely to object when someone pokes around their home and allows Mr. Gannon to inspect the hive.[33]

This narrative makes it seem that bees are apt to "mellow out" and not cause any potential trouble for neighbors if they are cared for by the right people. Good beekeepers can keep quiet, docile bees that do not need to be feared. Beekeeping is often described as a Zen-like practice, as exemplified in this statement from the same article: " 'I can't think of anything more relaxing than sitting in front of my beehive, drinking a beer, smoking a cigar, letting the bees fly,' Mr. Gannon said on a recent Saturday afternoon."[34]

Importantly, people had cultivated urban community gardens and had kept bees and other nontraditional animals as pets and livestock before the recent media coverage of these new trends. In immigrant enclaves and ghettoized and working-class neighborhoods, bees, chickens, and pigeons are kept as food and pets without much media attention and have been in this way for centuries. Like water towers, pigeon coops in particular are signs of an urban skyline, making appearances in many films as rooftop scenery or someone's urban hobby. As a cultural endeavor, breeding and flying pigeons in particular has been a way for white ethnic and nonwhite ethnic blue-collar people to connect with the environment and forge social connections. As the sociologist Colin Jerolmack shows in his ethnographic study of pigeon flyers in NYC, keeping pigeons cultivates solidarity across Italian, Hispanic, and African American men of different ages.[35] Pigeons and their human caretakers exist alongside the new wave of beekeepers and urban farmers, but they are largely invisible in the media. The demographic difference between these groups splinters them socioeconomically along educational, ethnic, and class lines.

The changing social demographics of beekeepers are exemplified in the article "Urban Gardener: Generating Buzz and Honey," published in the *Wall Street Journal* in September 2010, which reports how the urban elite is attracted to bees.[36] In this case, York Prep, a private school on the Upper West Side, is raising 500,000 honeybees successfully on its rooftop apiary. Bees live uptown and they also dwell in expensive neighborhoods that serve the culturally well-heeled. One of our informants, who tends a hive in Manhattan's Chelsea neighborhood, which is home to most of the city's art galleries, is now caring for bees housed on the roof of the Whitney Museum of Art. Given the fact that artists have been riveted by bees and have used their honey and wax to make artwork for centuries, it's not a stretch to imagine how bees could gain a temporary residency at a venerated arts institution.

However, beehives have been set up in less idyllic spots. Figure 3.4 was taken at Eagle Street Farms, a rooftop garden and apiary in Greenpoint, Brooklyn, also home to waste-treatment plants and superfund sites. In the photo the urban farmers and beekeepers Annie Novak and Meg Paska are inspecting frames containing bees and honey. Rooftop farms, community-based farm programs, and the introduction of chickens and

Figure 3.4. Annie Novak (left) and Meg Paska (right) on the Eagle Street Farms rooftop with bee frames. (Photo credit: Mary Kosut)

bees are tied to larger cultural trends, personal lifestyles, and philosophical perspectives that involve cultivating and integrating ecopolitics into everyday urban life. Novak and Paska are examples of a demographically new breed of human cultivators (sometimes referred to as "hipsters" in the media). Educated, relatively financial stable, racially white, and often female, this cohort lends credibility and interest to a phenomenon that isn't necessarily new. That is, immigrant enclaves and ghettoized communities in New York City have looked after bees, chickens, pigeons, and community gardens before bees had flown onto our larger cultural radar or Al Gore's *An Inconvenient Truth* popularized an environmental crisis.

When beekeeping became legal in New York City, many keepers breathed a sigh of relief, but at the same time some voiced a concern that not just anyone should buy a queen and a box of bees to put on a fire escape. Because bees generally fly in about a two-mile radius from their hive, neighboring bees intermingle and transmit not only semen between each other but also diseases and mites. One hive can affect another, and there is much concern over ill-prepared neophyte beekeepers that may inadvertently contaminate a robust hive. Beekeepers are self-interested in the health of their own hives and their longevity. The idea that anyone can do it, learning as you go, underestimates the labor and knowledge involved in providing a viable home base for bees. Even though bees can be left alone for weeks, their needs are not as simple and straightforward as what a household dog or cat needs. As the sociologist Jen Wrye explains, dogs and cats are "highly intelligent, adaptable and readily trained," and as such "can fit in to humans' lives and dwellings, whatever shape they take, quite easily."[37] Honeybees are not trainable per se and they don't live indoors. Our interspecies interaction with bees is largely based on the bees' own terms and needs for survival, and this is one of the reasons why not everyone is cut out for beekeeping. Nonetheless, many new urban beekeepers believe it is worth the effort. Through tending and cultivating bees amid chaos, debris, grit, noise, and concrete, there exists a potential relationality and intimacy that some individuals say they are seeking. Beekeeping in the metropolis can be a powerful and transformative "natural" experience, a way for urban humans to commune with a nonhuman species.

The Hipster Beekeeper

The complex cultural and demographic makeup of cities has fueled major American subcultures like hip-hop, graffiti, and skateboarding. Often youth-driven, aesthetic-cultural subcultures such as these began on the periphery before those outside of local urban communities adopted them. In the words of one twentysomething New Yorker, urban beekeeping is just beginning to "blow up." Similarly, Kirk Anderson, founder of the L.A. beekeeping group Backwards Beekeeping, said that a few years ago there were only six people in the group, but as of July 2011 there were about ninety to one hundred people attending the monthly meetings.[38] Likewise, a *New York Times* article reported that the San Francisco Beekeepers' Association had roughly fifty members in 2000, but by 2010 there were at least four hundred Bay area residents tending bees.[39] The interest is echoed in New York as evidenced by a spike in local beekeeping groups and alternative spaces that offer beekeeping classes and lectures, such as the Brooklyn-based workspace and education center 3rd Ward, which houses working artist studios, offers fine art and craft courses, and holds gallery openings and multimedia performance events. In addition to a course on how to raise and keep bees, the center offers drawing, graphic design, and jewelry making classes and is in partnership with Etsy, an online atelier for crafters that is also based in Brooklyn. This spot draws a generation of beekeepers who are in their twenties and thirties, interested in the arts and a DIY aesthetic fueled by social networking. Local honey grown upstate has been a staple in farmers' markets throughout the city for over a decade, but beekeeping itself now has a DIY cachet for some. So, unlike in the 1970s, where bees were coming to get us, simultaneously vilified and exoticized, presently bees have become a cause célèbre and, for many, a "cool" thing.

The NYC Beekeeping Meetup group currently has 1,232 members, a discussion forum, and a calendar of activities. This meetup group hosts talks, gatherings, field trips, and workshops. It also operates a beekeeping class over the course of twelve consecutive weekends that trains its students to acquire the fundamental skills necessary for maintaining a hive. In 2009, the beekeepers' school graduated thirty students. In 2010, in the class we attended, there were three times as many students eager to learn the dos and don'ts of beekeeping. Not only were there record numbers in attendance, but there also was a change in the demographics of students. The classes were usually populated by equal numbers of men and women, and their ages varied from people in their early twenties to sixties. Notably, almost all members were white. There have been shifts in gender and age in beekeeping groups, but they still remain, for the most part, racially homogeneous. This may be due in part to the wider cultural interests that neophyte beekeepers share, such as local food movements like community supported agriculture (CSA), organic urban gardening, participation in food co-ops, and other aspects of green lifestyle and consumer movements. Most of these hobbies and activities require discretionary income and time, however, and tend to be socially driven.

Importantly, the Internet has provided a way for those who may be isolated or disconnected from each other to form virtual communities, largely through websites and personal blogs. Content runs the gamut from sharing general friendly how-to advice (what tools you should buy, how to assess hive health) to more politically and philosophically driven beekeeping practices. Not surprisingly, the diversification of the larger beekeeping community has led to the splintering of groups and territorial arguments. The era of what one elder NYC beekeeper referred to as "the cool hipster blogger" beekeeper has begun. What exactly is a hipster beekeeper?

Like 1960s hippies and 1970s punks, hipsters are a contemporary subculture of young people linked to bohemianism and a particular identity and lifestyle. Hipsters are typically situated within urban neighborhoods and defined by their interest in alternative consumption. Hipster culture has been the subject of books such as *The Hipster Handbook* and *The Field Guide to the Urban Hipster.* Their identities are grounded in a recognizable style of dress, indie musical interests, and an ethos that is purposefully antimainstream politically and aesthetically. They tend to be white, well educated, from middle-class backgrounds, ironically fashionable, and noticeably thin—they have a certain look. Hipsters have clearly been differentiated and targeted by marketers, from American Apparel's Polaroid-esque pornified t-shirt ads to the emaciated silhouettes that embody iPod campaigns across America. Much like the punk scene sociologist and media theorist Dick Hebdige chronicled in *Subculture: The Meaning of Style*, the commodification of all things hipster has resulted in a shift in cultural meanings and demarcation of the archetype.[40] Similar to other styles that came out of human ingenuity (i.e., punk or hip-hop), there is a capitalist push to co-opt the style and mass-produce it—neutralizing the original transgressive ideal. Once companies notice that hipsters can be marketed to, the identity itself can be bought, packaged, and sold.

Media definitions of the hipster have expanded, as illustrated in a 2007 *Time Out New York* article that asserts "the profile of the typical renter in the area is changing from the 'counter cultural' hipster to the 'more mainstream' hipster and young professional."[41] Hipster is a loaded term and can be used descriptively but also pejoratively. Hipsters might not use the term to describe themselves—actually, we doubt if any hipster would. Since "the hipster" has become visible in popular media and used to market a plethora of products, from Converse sneakers and Camel cigarettes to luxury condos on the Williamsburg, Brooklyn, waterfront, it is more hipster-ish to reject being identified as a hipster.[42] Rather, it is more of a label that is ascribed to certain young people than one they embrace. In this regard, no young urban beekeepers identified themselves as "hipsters," and we are cautious of using the term to describe them. While in the field we have heard the term "hipster beekeeper" used negatively to describe young beekeepers and to indicate beekeeping as a temporary fad, it is also a designation uncritically heralded in media stories. In other words, the urban hipster beekeeper is based on a real demographic of young people who share beliefs and practices, but it is also simultaneously a stereotype, a construct, and, at times, a critique.

So-called urban hipster beekeepers tend to be more green and environmentally friendly—concerned with buying craft beer and homemade bread, rather than pricey Ray-Ban sunglasses or $50 vintage t-shirts. Members of the new generation of bee enthusiasts that are lumped in with the rising hipster beekeeper class are linked by cultural interests and style, rather than financial security. They tend to be educated and urbane, but not necessarily economically well off. At least half of our informants had learned the basics of beginning beekeeping by reading books on the topic, but others also took free classes and courses that charged a fee. A few had experience with beekeeping in their childhoods, growing up mostly in the rural Northeast and Midwest. In these cases, beekeeping was a part of early childhood, being taught informally by family and friends. But almost everyone had to brush up and also make an investment—beehives may be inherited but they are in no way easily transported. Previous experience notwithstanding, most people have to start from scratch. Establishing a new hive is a venture that could easily cost upward of $500 for a box and frames, suit and veil, hive tools, and the actual bees themselves. In our research, we have heard of people buying specific types of queens for between $30 and $100 each. Beginning beekeeping is a serious investment, in addition to the time, effort, and energy it takes to set up and nurture a hive.

Throughout the course of our fieldwork, we asked beekeepers why they thought so many young people seemed to be gravitating toward urban homesteading, beekeeping, and working on collective city farms. Caroline, a twentysomething and also a new beekeeper, said, "It may have to do with the economy tanking. . . . People are cutting back." Her friend Gabriel agreed and added, "In times of great economic crises and even war, people want to take care of themselves. An example would be victory gardens." Victory gardens, also known as war gardens, were planted in the United States during World Wars I and II by private citizens as well as in public parks to bolster the national food supply in a time of collective crisis. Having some sort of control over their food source and lifestyle is appealing for many younger people today, particularly middle-class college graduates that have a degree of choice and some discretionary time and income. Of course, you can't always control how much you get paid or whether you get laid off, but you can try and keep bees alive, raise chickens, and have a garden. In the spirit of the Occupy Wall Street movement, some are employing a DIY resistance to the status quo. In this case, urban homesteading is a means to reclaim what we have forgotten as a culture—to provide basic things for ourselves rather than purchase them. It's a small-scale and intimate effort to take charge of the space around them.

There is also an attractive and clearly oppositional component about this trend in terms of living in a major global city and being able to say that you farm and keep chickens and bees. There is cachet in keeping bees—the equipment, the sting, the historic and cultural mystery surrounding the hive. And it's a more tangible way to engage with nature and animals beyond a walk in the park or caring for a cat. Cerise, a beekeeper, reflected on the new generation of urban farmers and beekeepers:

> I think there is a few generations of being lost—not having a connection to land or to a sense of things, cycles. I think that is a fundamental need. I think that skill-based craft has been missing for a few generations. I think that people are putting value in that and experimentation is also a draw for beekeeping. I think beekeeping is an entry point; you don't have

access to land, poor soil or you are in the city means you can't do much. You can keep chickens and you can keep bees. You can brew beer; you can bake bread. That's about it.

If there are such things as hipster beekeepers, they are best characterized as "green" hipsters. They tend to be interested not in consumption per se, but in producing things for themselves—whether it's their own honey or knitting their own sweaters. It's a subculture of young people who think it's cool to make things, to bring the rural into the city, and to resist dominant patterns of food production. Unlike scientific beekeepers, who tend to be older, hipster beekeepers are typically more in line with backwards approaches of minimal treatment. The geographic and philosophical overlap of so-called hipsters with communities of artists or those who practice alternative health such as homeopathy has also encouraged collaborative projects with bees in art and health. Hipsters can use beeswax and pollen in their art projects or create tinctures of bee propolis for wound care. Through their rejection of mass-produced goods and a turn toward DIY production, they are very drawn to the bee as a powerhouse of productivity. There is a clear connection between being able to make things with bees and the performance of the beekeeper.

The exchanges between humans and honeybees are always relational; every hive functions by and large on its own terms, but each one is transformed when it is invested with particular human meanings. These meanings may arise from green practices, holistic philosophies, and DIY approaches, or scientific paradigms and procedures, and in some cases the demands of the marketplace and the goal of profit. Honeybees sustain human endeavors as part of a taxonomy of living objects to be studied and classified, and they are agents of pollination in agricultural systems. In these ways honeybees are written into the larger cultural imagination as a sanctioned and useful species: one that we work with and that works with us, albeit under particular human terms, ideological frames, and landscapes.

Save the Queen: Consumerism and Ecopolitics

Domestic beekeeping necessitates that we, to varying degrees, adapt to bees, not vice versa. In this context, it's interesting to consider whether beekeeping is some form of atonement for environmental degradation perpetrated by humans. Or perhaps it is a way of giving back and getting the wasteful and toxic sins (of our species) forgiven through this seemingly organic and selfless act of caring for a threatened species. Within this frame, bees have become ambassador insects to advertise green consumerism, the environmental movement, and the ethics of extinction.

Regardless of age or socioeconomic status, arguably, many would agree that we are living in an age of environmental degradation and that there should be steps taken toward sustainability. This process could encompass more environmentally responsible practices taken by corporations as well as an implicit and sometimes explicit ethical responsibility on the individual to live and consume in a way that reduces (or at least acknowledges) our ecological footprints. This green cultural climate affords an ideal time for us to pay attention to the plight of the bees. In countless newspaper and magazine articles and television reports, we hear the mantra that bees matter. Because they are often illustrated as adorable and fuzzy, bees function as a perfect poster insect—a ready-made logo—for green consumerism.

Green consumerism is difficult to neatly define, because it is a sphere of sundry products and practices that intersect with capitalism, environmental ideas, and individual consumption. Some products are considered green because they are made locally or under conditions of fair trade, yet "green" can also be used to label the way that people consume such as reducing carbon footprints or engaging in dumpster diving. Green consumerism has been fueled by increased public awareness of global warming (in some schools a part of elementary, high school, and college curricula), the institution of recycling programs in major cities, and general narratives that place individuals' everyday lives within a planetary ecological crisis. Green consumerism is founded on an ethical impulse that reaches beyond the individual. We are encouraged to buy products that can potentially help to decrease pollution, deforestation, the emission of carbon dioxide in the air, and so forth. Buying into green consumerism may be altruistic for some, but it is certainly fashionable and trendy. Green consumerism, like many kinds of consumption, is connected to self-identity, a fact not lost on marketers and corporations. Part of contemporary urbanite conspicuous consumption is the clearly visible low-impact display of reusable sacks, aluminum water bottles, and hybrid cars.

Green consumerism is sold to us by companies; we may not necessarily have to "go green" because we can simply buy green. As discussed, Häagen-Dazs ice cream has adopted the bees as a cause. The company constructed a website devoted to the preservation of bees and CCD. It's labeled helpthehoneybees.com, and its logo is "HD hearts HB," which stands for *Häagen-Dazs loves honeybees*. The website explains the honeybee crises, why we should care, and why the corporation cares: "At Häagen-Dazs ice cream, we use only all-natural ingredients in our recipes. Bee pollination is essential for ingredients in nearly 50 percent of our all-natural superpremium flavors. Our goal is to raise awareness of the honeybee issue so that our communities work together to bring them back."[43]

In addition to the ecofriendly manifesto on its website, Häagen-Dazs has also donated funds to both Pennsylvania State University and the University of California at Davis where it hosts the Häagen-Dazs Honey Bee Haven within a "pollinator paradise demonstration garden." The relationship between UC Davis and Häagen-Dazs is what the sociologist George Ritzer calls a type of "implosion" between consumer culture and educational institutions. In *Enchanting a Disenchanted World*, Ritzer argues that contemporary consumer culture is characterized by the erosion of boundaries among schools, churches, and medical institutions, so that these (among other areas of life) become "landscapes of consumption" that are often "disneyfied."[44] Here honeybee extinction becomes part of a landscape of green consumption.

In addition to buying expensive ice cream, consumers are also encouraged to buy t-shirts to support the cause (a portion of the profits go to bee research). The Häagen-Dazs site redirects to another site that sells not only the company's bee-saving logo but also t-shirts that say in bold yellow and black letters "Long Live the Queen," "Save Our Busy Bees," and "Keep the Hive Alive." If you are unable to cultivate your own bee garden, or keep a hive on your roof, you can buy a t-shirt. The purchasing of certain products, even if it is a t-shirt with a green message, is one way in which people can play a part in the green consumer trend and feel as if they are participating in a larger cause. As the geographer Juliana Mansvelt notes, it is not uncommon for green advertising such as this to contain an educational message, one that "informs the consumer about the added value of using green products, especially for the community and future generations."[45] Whether buying ice cream made in part by bees, or advertising their extinction through fashion, consumers have an opportunity to align themselves ideologically with ethical consumption. Of course, supporting green as an idea is much easier to do than radically changing your consumption habits. Reducing your global footprint cannot be accomplished through buying a pair of Tom's ecofriendly shoes or a bag with a bee picture on it.

Endangered Bees and the Environmental Movement

Bees are not the first nonhuman species to be placed on the list to be saved from extinction or to occupy a regular place on the cultural radar. Technically, bees are not under threat of dying off as a species at the moment, but the notion that they *could* be is implied in the media. Long before bees gained popularity and press, the public imagination was stirred by images and narratives of dolphins, whales, and penguins in peril. One of the first large-scale promotional crusades for an endangered species was the Save the Whale campaign of the 1970s. Aptly, it has been referred to as the "great granddaddy" issue within the broader burgeoning environmental movement that started more than thirty years ago.[46] In the past, photos of a few adventurous Greenpeace activists in tiny motorboats battling an enormous whaling ship were viewed by most of the mainstream public either as too radical or as well meaning but foolhardy. What's different today is that environmentalists have convincingly linked larger eco-issues such as pollution, oil spills, and ozone depletion as affecting not just the whales but the planet as a whole. The problems of whales (or of polar bears or tigers) and of humans are one in the same.

The World Wildlife Fund and other international environmental groups have long understood that certain species are more "charismatic." Not all animals function as endangered poster children. Land

and sea mammals with large eyes and/or humanlike faces are particularly compelling to humans. According to the sociologist Jen Wrye, much like the appeal of pet animals, there are certain "neonatal physical characteristics," such as "proportionally large heads, big circular eyes, and soft fur," that facilitate human emotional relationships (particularly between children and parents) that are transferable to animals.[47] We can presumably look through their eyes, and into their faces, ostensibly communicating on an interspecies level, regardless of whether we actually understand one another. Some animals, like dolphins or elephants, are possibly even capable of being sentient, stirring empathy and feelings. These animals in particular have also been shown to be intelligent by human standards, possibly possessing self-awareness and, in the case of elephants, long-term memory. Ultimately, the equation is straightforward; if we feel something for a like-minded and emotional other, we are more likely to care. On some level this translates into a moral imperative and varying degrees of action. At the very minimum, there is a large-scale cultural awareness that certain species are at risk and their existence (or extinction) may reflect directly or indirectly upon our own. The World Wildlife Fund, Sierra Club, and other conservation groups have done a successful job advancing the notion that we are all in this together, whether mammal, human, or insect.

In order to promote animal extinction as a salient human issue, and to communicate information as simply and clearly as possible for those who are not biologists or environmental scientists, the World Wildlife Fund categorizes endangered animals into four categories: flagship, keystone, priority, and indicator species. Flagship and priority species are the most recognizable and familiar poster animals that must be saved. The former represent broad environmental regions and issues, and the latter are used to hone in on a particular habitat in crises. Standard flagship species are whales and giant pandas, while penguins, monarch butterflies, and brown bears are representative of particular regions under ecological stress. Under these definitions, bees don't function as poster animals, most likely because they are too small and populate too many diverse regions. Penguins live on only certain continents: they can't survive in Italy or Qatar without the aid of humans. But honeybees can and do inhabit and thrive across countries and continents. They are native to Europe, Asia, and the Middle East and have naturalized in North and South America as a result of colonization. The only continent they are not naturalized in is Antarctica. Bees manage to live in highly diverse geographical locales, from the equator to Northern Canada.

Notwithstanding, bees fit under the classifications of keystone and indicator species. According to the World Wildlife Fund, "a keystone species is a species that plays an essential role in the structure, functioning or productivity of a habitat or ecosystem at a defined level (habitat, soil, seed dispersal, etc.)."[48] Pollination clearly lands under this definition. Indicator species are useful for creating empirically driven arguments that resonate with a sometimes-skeptical public. They are layman's proof that a habitat has undergone an observable change and is in danger because of it. If we conceptualize bees as an indicator species, then we must ask ourselves what the disappearance and death of the bees point to. Specifically, what change in the environment has caused stress for bees? This is a salient question for beekeepers and entomologists. Unsurprisingly, given their disparate agendas and working relationships with bees, each has offered divergent answers. When we asked one young but highly experienced beekeeper if bees were the new dolphins or whales, he looked at us incredulously and said that "bees don't need saving." Another beekeeper argued that bees were "co-opted" by greedy capitalists as "a hood ornament for the Mercedes-Benz they hope to buy as a result of all the fundraising they can do" by using the phrase "help us save the bees." Both of these beekeepers were skeptical of human attempts to save the bees, as in these cases they are believed to be motivated by either hubris or greed.

Bees have become increasingly popular, vulnerable, and iconic creatures because of CCD. Concern and controversy swirl around the insects with an ever-expanding explanatory network of causes for their disappearance. Unquestionably, CCD stories are easy to locate; it is much more difficult to find a consensus as to what CCD is, what caused it, and how it should be treated. There is so much emotion but also tension and disagreement around bees as a threatened species. In a metaphorical arms race, neophyte urban beekeepers cultivating hives are pitted against government/corporate-funded scientists developing monocrop pesticides. The environmental buzz around bees is both personal and political.

CCD provides a time for reconciliation, an opportunity for practicing the mutual regard necessary for species to live more sustainably. It strikes us that part of this buzz is a human reckoning with the sins of our species—pollution, global trade, chemical proliferation, exploitive labor practices—and that being green emerges at a time of atoning for these sins. A new generation of beekeepers has emerged in urban locales connected with the larger apparatus of environmental threat. Reparations are due to the bees if we are to coexist.

4

Being with Bees

Intimate Engagements between Humans and Insects

Worker bees can live from four to nine months during the winter, but the average life span of a worker bee in summer ranges only six to eight weeks. Aside from the queen, who may live to be several years old, bees' lives are fairly abbreviated when compared to other species. Elephants and certain parrots live up to seventy years, and queen termites can survive for fifty years. Notwithstanding, comparatively short-lived honeybees do things for human environmental survival. Bees buzz through the world quickly, but in that time they make a significant mark on our lives. On a large scale they assist in human survival and on a smaller intimate plane, they get under our skin. Honeybees affect people: they transfer pollen and create beeswax (useful processes and material objects), and bees trigger people to act, to run, to tremble, to smile, to yell, and to be still. In figure 4.1, a beekeeper admires the curious shapes of two sunlit combs.

Bees are not ordinary insects. Humans are wrapped up in what bees do, in part because they trigger our vulnerability: we want to repel and contain them. They are fascinating, but elusive, and for many of us, they are frightening. Their symbolic power is informed by the sting, a transmission of venom conjuring fear and anxiety and, in rare cases, death. We were surprised when we learned that the number of human deaths by bee stings is not as statistically significant as one would imagine. The myth of being stung to death outweighs the reality. According to a World Health Organization census report from 2000, there were fifty-four deaths caused by bees that year in the United States.[1] In the words of Jim Fischer, "Bees are safer than the bus. And when they outlaw buses, only then can they tell me that bees are dangerous." Even though bees are not naturally aggressive toward humans, one sting in a sensitive spot can be acutely painful.

Figure 4.1. A beekeeper inspecting a top bar comb in sunlight. (Photo credit: Mary Kosut)

A stinger is a weapon, but honey is nutrition. For these reasons, the bee is an assemblage of human desire and aversion, an organism of contradictions and human-bee intra-actions are fluid. Some people protect themselves from the bees to be safe, while at the same time they feel it is their responsibility to protect the bees as their self-appointed stewards. We struggle with the bee in so many ways, but ultimately they are out of our control.

Humans have intimate relationships with other species, but they usually involve mammals, especially dogs, cats, horses, and sometimes whales, dolphins, and even bears.[2] Bees, unlike mammals, or even reptiles, are classified as invertebrates. Invertebrates are animals that don't have spinal columns; the group includes insects, spiders, scorpions, worms, leeches, mussels, snails, and many others and make up 97 percent of all animal species on earth.[3] Even though they dominate the animal kingdom (at least in numbers) we are often at a distance from most invertebrates. Perhaps this is because many do not have human-like faces such as mammals. A clam, like a worm, doesn't even have a discernible head. For this and other reasons humans are more used to crossing the literal and theoretical borders of the animal/human binary with certain species, not usually those without backbones, especially creatures we commonly think of as *bugs*.

Bugs are everywhere in our backyards, our beds, and our bodies; they thrive underneath and on top of the skin, gorging on our ample fleshy surfaces. In the words of *Discover* magazine each of our bodies is a planet, a teaming ecosystem that contains ninety million microbes, including many types of bugs like "minuscule, eight-legged Demodex mites [that] nestle head down inside the follicles of the eyelashes, feasting unnoticed on skin cells."[4] But not all of these invisible bugs are harmful or parasitic, even the ones that live off of our eyes. Humans live harmoniously with most of them, most of the time. To paraphrase the political theorist Jane Bennett, bugs *are* us because "the *its* out number the *wes*." We are not simply embodied as humans; we are "an array of bodies, many different kinds of them in a nested set of microbiomes."[5] But it is hard for us to fathom the sheer numbers and kinds of bugs that make up the body's ecosystem, let alone the volume and diversity of insect species that visibly populate the world. It is sometimes difficult to see that we share and make up common or overlapping worlds. This inability to see bugs as always and everywhere present in a shared world is a form of human privilege. Humans treat themselves as a special class in the hierarchy of beings.

Although most of us have experienced running from, squishing, or simply ignoring most insects, there has been something of a bug renaissance in the social sciences. Scholars that study humans are moving toward, rather than away from, bugs.[6] The anthropologist Eben Kirksey calls this attraction "insect love," a passionate scholarly attention to and reverence for these maligned species, with an emphasis on our intersections.[7] Another anthropologist, Hugh Raffles, author of the award-winning and popular book *Insectopedia* (2010), epitomizes insect love by paying homage to our conflicting, intense and sometimes breathtaking interspecies relations. Raffles explores the vast continuum of insect/human entanglements and how we are stirred by the smallest of creatures in radically different ways, whether through getting assaulted by malarial mosquitoes in the Amazon, betting on Chinese cricket fights, or obsessively illustrating the mutations in leaf bugs.

Kirksey's own "unrequited" insect love was sparked by a species of ant, *Ectatomma ruidum*, which he came to know through field observations in Panama. In the process of observing their networks within and between nests, Kirksey found that the ants did not act as autonomous members of distinct "superorganisms," nor were they inherently destructive creatures as previously described. Instead he thought that *Ectatomma* colonies should be understood as "ensembles of individuals" that are enmeshed with "other beings through relations of reciprocity, accountability, as well as kinship." The ant's relationship with each other, other nests of ants, and the human-marked landscapes shared with us (and countless other species of flora and fauna) calls for a more complicated way of considering the lives of other species. There is much to be gleaned from looking more carefully at these ants, so too, crickets, leaf bugs, and honeybees. Whether we like it or not, we are entangled with these "othered" creatures, and as Kirksey suggests, humans should learn to "better embrace species like *Ectatomma*, cosmopolitical amphibians that are good for humans to live with in common worlds."[8]

This chapter explores the intimate common worlds of humans and bees and what happens when we come into close contact with each other. What happens when bees permeate our skin and our consciousness? Beekeepers feel a buzz, a slight intoxication, enthusiasm, and exhilaration in the presence of these insects. This feeling is what we term the affective buzz, a transformation through bonding with bees. The affective buzz is a form of insect love, and also, similar to other types of love, it encompasses fear. In the case of bees, it is fear expressed in a sort of twitchy, jittery, and jerky flailing around. We have emotional entanglements that are simultaneously positive and negative as bees can provoke a range of feelings from pain and panic to awe, wonder, and calm reflection.

The exchanges between humans and honeybees are also animated by literal buzzings, so we also attend to the real buzz of bees as an embodied sound. The buzz emanates from bees' bodies, to the beekeepers and our own; it bounces back and forth in a jumble of physical sensations and intense emotive states. The embodied and affective buzz between humans and bees is fairly easy to pin down from a human perspective; we narrate how the bees make us feel and we hear the sounds. The bees' buzz speeds people up and it slows them down. Like some form of insect drug, bees have a physiological effect on the body, affecting the way we think, act, and move. But how do these entanglements affect the bees?

We are aware that we must cross many borders and spaces to get to the place of the bee. With this in mind, we are interested in discovering animal/insect gestures. For us, these gestures include signs and acts that are not necessarily reciprocal, or fixed, but are emergent and in motion. They are interspecies echoes and reverberations—a cacophony of sounds that are made, and heard, but not necessarily understood by either insects or humans. The affective buzz is not unidirectional but is what Jane Bennett refers to as an "agentic assemblage" made between two actants: lively human and nonhuman entities.[9] In her book *Vibrant Matter* (2010), Bennett builds on the ideas of Baruch Spinoza, Gilles Deleuze, and other philosophers, seeking to complicate how we consider "life" and "matter." She theorizes a "vital materiality," which erodes the distinctions between humans and nonhumans. In this frame, we follow the vibrant vital buzz ricocheting through the bodies of bees and people, two species striving to coexist in common worlds.

The Work of the Beekeeper

I'm trudging ankle deep in muck, circling around a barn, next to a pond with croaking frogs (unbelievable cliché unfolding), and suspicious horses that are glaring at me with too-large eyes; go back to Brooklyn. Boots tucked into pants so that ticks don't invade my body's cracks. Also been warned of poison ivy. Inevitable bee stings. My first bee yard visit, and I am walking alone up a hill behind a farm toward tens of thousands of bees that have been jarred, split up, and forcefully shaken from their whirring but previously still lives—hives. Makes me dizzy, I am thrilled at the feral-ness and the foreignness of it. Feels like when you are young and scared but you have faith in the adults—like here ya go kid, you can feed the alligators but you have to move slowly. The bait is at the end of a stick, but it doesn't seem quite long enough. What am I getting into? Embodied fieldwork. Bee-ing there. Trust. I have to trust my human subjects, and give over to the stinging insect as subject. I am not in the space of metaphor or anthropomorphism; I am in the space of the bee.

I begin to hear voices and see makeshift boxes, crooked wood and metal structures. These are the beehives, and like old buildings, you wonder if they are strong enough to house anything. People are working the bowed hives, bent over. Lisa Jean has a white veil on and is bent over and helping split up hives, looking for queens, eyeing up any signs of life—both good and bad. The good signs are a full even bar loaded with honey and nectar and brood . . . some of the top bars are a bit broken and uneven and break and fissure a bit when she pulls a bar up from the boxes. She's moving purposefully and finds a Queen somehow amidst a full healthy frame of bees that is overflowing with larval gooey activities. Great for her but I'm going to have to touch all this mess without gloves, veil, suit. Fuck. Within ten minutes I am sweaty, coated with dirt and starting to itch, just from looking, I haven't even moved toward the bees yet, just away from them as they breach the space around my body. Out of my league/space.

Bees are looping around in circles—some rapidly, others more crazed, they are like toddlers, flying in the air and they cross and crisscross the hives and the trees and us and make the air sometimes clogged with zigzagging. Try not to focus on all of the careening and I can imagine bees accidentally banging head-on into each other and dying in midair. I learn that long hair is not a good idea as bees can get caught in it, become angry and confused, and sting. Sam lends me his hat and I feel a bit more protected when I shove my hair into it—maybe he sees that I am nervous and tells me to talk with Lisa Jean in the shade. Guess they are more calm in the shade or I become more of a muted target. Who knows. I want a veil. Somehow, miraculously, I am not getting stung. Watch bees land softly on my wrist, see them on my legs, shoulder, feel them bump into my head (the rapidly careening ones) and even had one bounce straight off my nose. Remain pretty calm, I think (except later when the roach crawled off the frame and up my arm). Others back a few feet away when the bees take over the local airspace. As I get updated by Lisa Jean I see her swollen lip, and hands and I see in her eyes that she has been stung hard and it hurts.

—Fieldnotes from Beekeeping Bootcamp,
"Farmer Jen's" bee yards, May 2011

Being with Bees

The first few times in the field, working with and around bees, we grappled with the obvious question of why people would do it. To be blunt, particularly in the beginning, we were sometimes anxious and on occasion came home with bee-induced injuries. As Mary's fieldnotes above state, it took some time before we began to get it—to see the bees through our informants' eyes. One backwards beekeeper told us that "people that keep bees don't follow a straight path—they veer off a bit. It takes a certain kind of person." We did find some commonalities between beekeepers: they seem to have come from some level of ecological awareness through gardening, food politics, urban farming; some have generational connections to beekeeping; and importantly, almost all of them talk about the practice being meditative and peaceful. Often the practices of beekeeping are a way to calm themselves and focus on something other than their "real" lives, because beekeeping requires that humans be present in body and mind.

The act of beekeeping is often deliberate and careful, as humans must adjust their behaviors to the perceived needs and sensibilities of the hive. Beekeepers are responsible for knowing when to do a hive check and when not to, when to give bees some sugar water, when to leave more frames in the hive, and when to split the hive to prevent the swarm. Ultimately, these choices rest in human hands. With humans-know-best hubris, we act to help our hives because we believe the bees might not be able to help themselves.

Likewise, it is also a carnal and sensate experience; the hives themselves are animated with lush sounds and smells. Working with bees involves both physical labor and levels of emotional labor that are cultivated over time as part of nurturing and maintaining a colony. We discovered that beekeepers often become intimately invested with their hives, and in particular their queens, because they are the key to a healthy productive colony. A weak or old queen cannot sustain a hive, so checking for her becomes a separate and meaningful act within the larger ritualistic process of tending and working with and for bees. But the experiential and embodied knowledge of beekeeping involves more than becoming adept at visual identification of certain signs of healthiness, frailty, or disease. There are corporeal cues (auditory, olfactory, visual) and intimate moments where beekeepers intersect with the hive, as not only vulnerable but also sensual bodies.

The Austrian social reformer and spiritual philosopher Rudolf Steiner, who advanced "anthroposophy," or spiritual science, wrote passionately about bees. He conceptualized bees as blood cells swimming about within a body, and embryonic and cellular development as a link between human and bee. Steiner asserts, "It happens the same way in the beehive, the only difference being that a human being builds up a body that only seems to belong to an individual, and the bee also constructs a body: the honeycomb, the cells."[10] The metaphor of a hive likened to the human body, wherein each being is made of separate but interrelated parts and systems, is repeated by theorists, entomologists, organizational psychologists, generals, ministers, and lay beekeepers. The hive, like the body, looks fairly contained from the outside, but a closer look reveals modest openings. These openings, whether holes, cracks, or orifices, are portals where objects and things move in and out and can come and go. For humans, and bees, the constant entering and exiting of hive/body is necessary for the consumption of food, waste removal, sexual reproduction, and physical exchanges within and sometimes outside of a species. One of our informants told us that putting his hand in the hive is "like putting your hand inside a body cavity." It was alive—throbbing, warm, and vibrating.

Being with the bees involves smelling, hearing, tasting, and feeling them through your own body. As one beekeeper explains, beekeeping can be focused and measured: "There is something very Zen about the process of taking care of the bees. You have to slow down and really pay attention because you don't want to upset them and you don't want to get stung. When inspecting a hive, your mind can't be anywhere else: it has to be on the hive observing exactly what is in front of you." The act of beekeeping involves perception—a responsive performance of mind/body and bee. Much like dance or practicing tai chi, there is a beauty in the movement and flows—choreographed and improvised at the same time. The performative nature of beekeeping also calls for embodied learning and sensitivity. Here the bee becomes the educator/teacher through a commingling and penetration of the senses. Becoming attached and in sync with a colony or a hive is a ritualistic process, but it is also a sensual one where insects and humans connect, overlap, and collide. Some of these collisions—the sting—are unpleasant and downright painful, but others are fragrant, delightful, therapeutic, and delicious.

When we went on our first hive inspections we were acutely aware of the possibility of being stung. This awareness was an obvious and ever-present embodied and emotional concern. However, part of learning to work with bees and observing the labor of others involves not only learning what to look for but also studying how the bees smell, sound, and sometimes taste. It was not simply an act of observing human-bee exchanges, but making contact with the bees through our own bodies. Mary's fieldnotes recall how one of her first encounters with honeybees was surprisingly sensual, specifically, the intoxicating feral smell of the bee:

> *The wax combs sometimes smell a bit like lemon. Lemony and faint. I don't have the knowledge of how to smell bees yet. When Sam opened the Langstroth or traditional box there was this thick sweet odor and I smelled it right away—it was sexy—smelled a little fetid or rotten, but pleasant. Like the crotches of people you lust after. My brain and olfactory senses didn't know what to make of it. I ask—"Does anyone smell that?" and someone said, "That's the bee pheromones." "They are smelling sexy." Asked Sam and others if they got really "close" to the bees and if there was any interspecies sexiness going on. Like how close can you really get to the bees?*

Other informants also talked about how the scent of the hive drew them physically closer to the bees' rhythms and sensibilities. B.J., a Brooklyn backyard beekeeper and chicken caretaker, explicitly talked about the seductive nature of the smell, likening it to fragrant truffle oil and human pheromones: "I love the smell. Like if I am weeding and near the side of the hive and I get a whiff of it and I am like oh my god, it is like truffle oil or something. It gets right here and you are like, whoa, lovely. And you know they are working. And it is oh, so like um, a heady good sex smell."

Meg Paska also reflected on the pleasure of smelling the bees, like B.J. above, and she acknowledged the significance of the smell as a sign of content and healthy bees, as "it usually indicates colony health." For

Paska, it is an aroma that similarly calls to mind human intimacies and memories. She described it as "a slight fermentation scent, very subtle. The bees' wax and propolis. For me, it is the most intoxicating smell I can think of. In the wintertime I find myself craving that smell. It is kind of comforting, like home or something. I just want to put my face in it." These narratives unearth the embodied nature of beekeeping and also the levels of physical exchange through the labor of beekeeping. In this case, the particular scents that permeate the hive space bring the bees closer to us. Some beekeepers inhale the hive and its energy and emissions. This sensual experience is another way that bees permeate human skin, triggering an emotional response through the colony's fragrance.

In addition to engaging the sense of smell as both a sign of bee health and/or pleasurable human stimulation, beekeepers are also attuned to the sounds of their hives. Numerous beekeepers confirmed that they try and determine what the bees might need based on their buzz. A hive can produce thunderous clatter as if the bees are yelling, but other times the bees' collective sound is more like a whir or hum. Beekeepers listen carefully to the buzzing to assess what they may be feeling or trying to communicate. At a hive inspection on Manhattan's Randall's Island, Jim Fischer explained how the frustrating constant roar of metropolitan life drowned the sound of the hives: "Up at Van Cortlandt, we would have been able to hear the bees much better, and I would have shown you the G-natural or G-sharp of a happy hive versus the A-sharp of a pissed-off hive—we can't hear much of anything on the 5-Boro building roof." The bees are able to effectively communicate with each other, but the din of a metropolis can sometimes impede certain embodied aspects of beekeeping, in this case listening to bees over the clamor of traffic on the Triboro Bridge. This is part of the challenges of urban beekeeping.

B.J. spoke about how she listens to the bees and what their sounds might mean: "It gets higher and lower based on their activity. When they are angry it gets higher pitched and louder. If I am in there too long, they give me signs. They can be smoked again, or get out and leave them alone." Similarly, Eric Rochow, a beekeeper and creator and producer of Gardenfork.tv, a videoblog, said, "You can tell when they are mad by the sound changing. It gets louder. When you open it up they are kind of looking out at you. Then when something agitates them the tone changes and when you smoke them the tone changes right away. It gets busier—you can hear this ring—they are sending a signal out. And then when they are bonking you it is a much higher-pitched sound. They are warning you and trying to get you to go away." Humans interpret bee sounds as if they are auditory codes to decipher insect behavior, but there is also an energy in the bee buzz that, like music, fills the body. You hear the bees not only in your ears but also as sonic vibrations reverberating throughout your arms, legs, and hands.

While the different pitches, notes, and levels of buzzing can sometimes be alarming to human ears, the bees' noise may also be meditative. It depends on the bees, the person, the site, the weather, and the moment. In the fieldnotes below, Mary describes how "working" with bees involves tasting fresh honey stores in addition to being caught up in a cacophony of sounds:

> *All the splitting up and re-queening is the work that Sam does. His co-labor with and for the bees. Today he is working with an audience (us). He is patient and upbeat and genuinely excited to see an overflowing frame, or to stick his fingers in the honey or break off a piece of comb. He makes jokes—"taste this, this is 'awful!' " (jocular and ironic). And I shyly put my finger forward around some oozing clear honey literally being made now. I keep it in one cell of the comb because there are active bees, squirming, caked and chunky/clunky all around. Sam just goes in with confidence and ease and the bees keep doing their thing. Swirling and whirling—their buzzing coming closer, darting and arching by my face and ears, but I hear a distant hum when the bees and the boxes are opened up. When he forcefully shook some of the frames and the bees dropped to the bottom of a new home (wooden box), they get more vocal and move faster and even with people talking around me all I heard was the bees' noise. I say "it sounds like water." I think that's the best way to describe it. It is a uniform noise—as a whole—but if you listen closely there are ebbs and flows and the sound moves like water around and inside your ears.*

In addition to observing the movement, the numbers of insects, and the kinetic energy of the hive, we struggled to listen to bees. It was important to observe what they were doing, to see them, and also to hear them. We wondered exactly what we were hearing: was it noise, a directed communication from the bees—was the buzz the bees' way of speaking to us? Or are we irrelevant and just listening in on private collective chatter? Failed interpretation notwithstanding, after spending more time with beekeepers and their hives, whether on rooftops or in urban backyards flanked by concrete, we began to wonder if the buzz of the bees was addictive, like the repetitive and constant sound of waves crashing or birds chirping, or music that you play over and over because it stimulates, calms, or soothes you. White-noise machines are sold for this purpose and there are also mobile phone apps that replicate natural sounds to soothe in unnatural locations—such as urban bedrooms, dentists' offices, and spas. Marketed as tranquil devices, white-noise machines, wishing to transform their association with noise, are re-branded as sound conditioners. Bees are natural sound conditioners, particularly in the city, as they soften and neutralize the inorganic, clamorous racket.

When working with bees, the buzz is ever present, regardless of whether it occupies the background or overshadows conversations with the people kneeling next to you. Perhaps this sound is a way that bees temporarily and inadvertently move people to pay attention, to work slowly, to be present. As a beekeeper named Christine describes, a hive has certain energy or power over humans: "When my grandson was young, we would bring him near the hive and it would have a relaxing calming effect. Stop crying like that." She paused and then told us, "There is something in the air about them." Similarly Meg Paska experienced a calming effect from being with bees:

> There is something really meditative about the process of inspecting a beehive [see figure 4.2]. Obviously bees are venomous stinging insects and though they are docile most of the time they have to defend their food stores and their young so they can be defensive, so you sort of have to take your time and be very methodical and move slowly and really focus on what you are doing so you don't get stung. And for me that has kind of been an interesting form of meditation. It has been really great for someone like me whose mind is constantly jumping from thought to thought. It brings me to one focal point and keeps my mind clear for a while.

Figure 4.2. Meg Paska on a Brooklyn roof inspecting a frame.
(Photo credit: Mary Kosut)

For some people, the buzz of the colony is like a meditative drone that offers moments of peacefulness and escape from the clutter of the mind and the immediate surroundings.

The sight, sound, taste, and smell of bees trigger a way of movement. Beekeepers learn a cadence and a pace from the bees, a type of collaborative free-form dance; humans sidle up and partner with bees. This somatic interaction with the bee is a form of animal husbandry: the practice of humans breeding and caring for animals over time. Sometimes it takes a few seasons to understand how a specific hive works and what the bees seem to want. There is a learning curve with regards to perceiving bees and ascertaining what they need. Beekeepers must pay close attention to the bees, take direction from them, and generally appreciate the subtle differences in somatic cues/clues. At the height of the season, beekeepers perform weekly hive inspections, but when the work is done (after about two hours), they silently observe the bees rather than rush to the next social activity or chore. Urban beekeepers develop routines that enhance their aptitude at hearing bees—going slowly, being deliberate, breathing deeply, paying attention, and speaking in hushed tones.

Suiting Up and Smoking Them Out

Being with bees may unconsciously cause people to slow down, but beekeepers purposefully control their bees—they actually *make* them slow down. Beekeepers desire docility and calmness from bees, so in essence they drug them before suddenly opening the roof of a hive. Opening the hive must be quite a shock to bees: the sudden burst of sunlight, dramatic change in temperature, and the movement of the colony often lead to a frantic scurrying around among them. Using a variety of plants (like burlap or pine needles), beekeepers "calm" the bees with a smoker so that bees cannot smell each other's alarm pheromones and trigger the entire hive to attack the human predator. (A smoker is shown in figure 4.3.) From the human perspective, honeybees need to stay calm and confused so that beekeepers can work them. But bees have every reason to be agitated when their hives are opened, as this work may also involve the sacrifice of a few bees, through decapitation (the movement of frames), broken wings (the replacement of frames), and crushed bodies (the lining up of frames). All beekeepers use smoke because it presents an opportunity for them, a safer entry into an unexpected world.

Having a smoker is one of the basic equipment needs that most urban beekeepers require. But there are other tools and gear that are not necessities per se. Some beekeepers argue forcefully that certain equipment gets in the way of good practice, particularly gloves, which may make for rough or clumsy beekeeping, leading to the deaths of bees. Because you are likely to work slower and more carefully without gloves or other protection, you may be more aware of the bees' movements.

Figure 4.3. Tim O'Neil with a smoker. (Photo credit: Lisa Jean Moore)

But working without the gear involves taking a risk. It heightens the exchange for humans. On the other hand, using a beekeeping suit, veil, and gloves mitigates the human danger. While it is less risky to humans, it may be more dangerous to bees because the human risk of a sting is lessened and carelessness is not as costly. The politics of beekeeping reveals that there is also a prophylactics of beekeeping. Beekeepers disagree over how much protection is enough.

As discussed in the field narratives here, some informants wear certain gear, particularly veils so that their faces and heads are not vulnerable. Other people get fully suited up. Eric Rochow told us he is not interested in freestyle beekeeping: he completely protects himself with a suit and a veil. He couldn't relate to the natural or freestyle approach because he pragmatically said, "The stings hurt!" and it's not something that he seems to be particularly interested in. Rochow employs the prophylactic gear because of the risky nature of beekeeping: "It can turn bad fairly quickly. I might not wear gloves as now I have been stung enough that it doesn't bother me that much." Rochow doesn't relate to the "badass" aspect of getting stung and "taking it like a man." For him, wearing the equipment brings to mind the adage "It is like my dad is better than yours" since he has the right stuff for the job. Rochow shared that everyone he knows now wears a veil. Another beekeeper, B.J., was similarly practical because of her job:

> I do get pretty suited up because I am a singer and if I show up to a performance and I have a giant growth—they don't really want to see that on stage. So I do get suited up. I was thinking of going to the hive three doors down and it is very quiet over there. So I was like maybe I will just put on the jacket and not the legs. But as the season progresses and they get more and more food in there and they don't want you to take it, I like to err on the side of caution.

Being cautious seems reasonable and rational, but there are some beekeepers who advocate going bareback. While this is a slang term that humans use to describe having sex without a condom, the term conveys the allure of freeing yourself and the danger of exposing yourself at the same time. Like other forms of extreme body practices, from skydiving to being suspended from flesh hooks, bees offer a thrilling unpredictable experience of human surrender.[11] Going natural and unprotected can be exciting and certainly more perilous. As Gita explains, "I like to go without wearing my suit. . . . It is so much freer." Likewise, Christine told us that she likes to go without a suit if possible because "it's so awkward and bulky." Like those who engage in unprotected sex, there is a negotiation among desire, risk, and consequences.

Of all of our informants, Sam Comfort was the most unprotected of all the beekeepers we visited. In fact, he did not use any beekeeping prophylactics and didn't wear any shoes on his feet either, leaving another sensitive area of the body exposed and vulnerable. On one hive visit he got stung on the foot, as described in Mary's fieldnotes:

> *Everyone seems to be getting stung today and we are so naked out here. I can't believe Sam with his bare feet and when he got stung he just said in a nonplussed way, "Oh, yah, got another on my foot." And then that other unprotected guy was like, "Yeah, my arm is swelling a bit." What? It was*

> *twice the size of his other wrist and the skin was bright pink and angry—it looked itchy and painful. I said to him, "That looks like it really hurts!" He says, yeah, you are gonna get stung if you work with bees. The other guys echoed this as well. Expect to get stung, and when you do make a point to casually mention it, if at all. Above all, they said avoid the gloves. Made me think of no glove, no love.*

Later that day, Comfort told us that the problem with gloves is that they can make people be too rough with the bees. As a prophylactic for the human hand, gloves are more than a physical barrier. They can hinder a potentially intimate and porous bee/human interaction. The choice not to use any protection seems very much a part of an embodied approach to being with bees—an attempt at a more open and perhaps closer relationship.

Bee Love: Getting Attached to Insects

The ways that bees look, smell, work, sound, and sting attract and affect humans. Their inventive mode of production arouses imagination—early humans visually depicted bees in ancient cave art drawings. Our relationship with bees is well documented in a variety of cultural contexts, from references in the Bible and Koran to images of bees in engravings, woodcuts, and illustrations. Representations of bees and ancient "honey hunters" can be found in numerous petroglyphs located on cave walls throughout Europe, Africa, Asia, and Australia.[12] According to the entomologist Stephen Buchmann, Africa has the largest number of petroglyphs depicting prehistoric bee/human intersections:

> Paintings showing bees, hunters, and the ladders, torches and honey containers . . . have been found on rock walls from Algeria, Libya, and Morocco in northern Africa to Namibia, Botswana, and Zimbabwe in the South. Much of this art communicates a sense of respect, and even awe, our ancestors must have felt for the bees they robbed. Their ingenious construction, the honeycomb, is often depicted in extraordinarily accurate detail.[13]

Although we can only speculate as to what these symbols meant for early humans they appear across many continents and cultures. Cave painters may have respected bees as a food source, or been in "awe" of them as stated above; regardless, they have a cross-cultural presence in human history. Poets, musicians, and artists alike have been attracted to bees, curious about the architecture of their lives and their labor.

In 1923, Rudolf Steiner gave eight lectures on bees, highlighting the "unconscious wisdom" of the hive and what humans could learn from bees about themselves and the cosmos. He spoke of a Platonic "love life" that develops within and throughout the entire bee colony: "The whole beehive is permeated with life based on love. In many ways bees renounce love, and thereby this love develops within the entire beehive. You'll begin to understand the life of bees once you're clear about the fact that the bee lives as if it were in an atmosphere pervaded thoroughly by love."[14]

According to Steiner, the bee's love is connected to the plant's love life. When a bee pollinates a flower, it "sucks its nourishment" and "brings love life from the flowers into the beehive."[15] Thus, Steiner believed we must "study the life of bees from the standpoint of the soul."[16] Cast in this light, bees could provide us with a threshold of sorts, a way for humans to become more enlightened beings. Sam Comfort echoed this notion: "Having bees around is so healing. They are not rational. They don't follow the rules, they are like love."

Some humans have been changed by bees. They get attached through time spent laboring over bees, especially throughout the spring and summer months when bees are most active, getting to know the "personalities" of the hives, or just hanging around them. When the work of beekeeping is over, some people linger around these insects rather than rushing off to their human worlds. Eric Rochow told us, "When we are done we will have a beer and just watch them. . . . We will put them back together and have a sandwich and it is very cool to just sit and watch them. They shoot out of that landing board, starting out at fifteen miles an hour—that is like an aircraft carrier. Where our hives are, the neighbors will come over and watch them because it is just this fascinating nature thing" (see figure 4.4).

But for some humans bees become more than a close way to be in contact with the natural world; bees transcend the designation of live natural entertainment. Through working with bees, humans develop relations of interspecies intimacy, levels of attachment that transform

Figure 4.4. Bees taking off from the landing strip. (Photo credit: Lisa Jean Moore)

human life. Comfort explained his experience of transformation in straightforwardly psychological terms: "I hang out with insects all day long—my thought process is no longer linear. It takes a certain kind of person to do it. . . . I have a psychological connection with the bees—this is important." Of course, not all beekeepers spend more of their waking hours with insects rather than humans, and not all beekeepers are as singularly focused on being with bees as Comfort. However, humans form attachment to bees in ways that change them. In seeing the bees, they reflexively begin to see themselves as individuals within a larger ecological assemblage of actors.

Although each of our informants had different philosophies, styles, and attachments to their bees, everyone we interviewed expressed a reverence and respect for what bees do and how they do it. Meg Paska's narrative sums up some of the major themes that surfaced throughout our time in the field and why people get so swept up in beekeeping:

> I think I kind of understand why [people] are just blown away. I feel like for a long time, I had a lack of spirituality and this lack of God in my life in some way and I am not like a religious person at all, I don't subscribe

> to any sort of dogma. But after a while you start to feel a lack in your life. So when I started keeping bees, a light bulb went off and I felt sort of an immediate reverence for what goes on in the hive. Partially because it is still outside the realm of complete understanding. And I personally think that if there are magical creatures in the world, honeybees are those creatures. I mean everything they produce, everything within the beehive can be used for healing purposes. Even down to the venom down to the bodies of the bees themselves. They live in this world without destroying or taking from their environment. They actually give. So I think that is kind of a rare thing.

As this quote suggests, being with bees can be a humbling experience. The patterns of bees' behaviors are observable, but there is something unknowable, even disconcerting, about the species. It's almost as if the bee is too good in human terms; they are hard workers, unselfishly producing healing agents for humans and sacrificing themselves for the benefit of the hive. In this way, bees are not only model insects—they are insect martyrs. People are attracted to the bee's sacrificing nature, a symbol of ultimate altruism.

Bees' awe-inspiring qualities sometimes give people pause to reconsider the imbalance of the human/insect divide and the notion that we have a certain degree of power over aspects of the natural world. As a beekeeper named Zan states, "Notions that we are superior to them is silly—the bees have been doing it without you. All you are trying to get them to do is to stay in the box. . . . They are really amazing creatures—you are the interloper in the whole thing." Many people considered the idea that humans may need bees more than they actually need us for survival. Those who are intimate with bees are struck by their capacities and their perceived competence as a species. As Rochow told us, "I have a deep, deep respect for them. The animal the size of your thumb is doing amazing things."

As human "interlopers," caretakers of beehives, there is recognition of the responsibility inherent in keeping bees alive. Beekeepers not only physically but also emotionally labor over their bees. When hives die, beekeepers are affected—they feel sadness, grief, and disappointment. Even for those who have said that they are less attached to bees, death is a way that they may become more emotionally linked. For example,

Caroline, a touring musician in her twenties, part of the new generation of urban beekeepers living in Brooklyn, explained to us that she got started by reading books in the library while living in Providence, Rhode Island, and then asked her friends to help her fund a new hive. She got twelve friends to donate $20 each and then she purchased the hives and equipment, as part of a DIY collective ethos. Notwithstanding, she clearly sees the bees as her responsibility. Through her emotional labor, her relationship with her bees became fairly intimate early on. She told us that after being on a European tour for four weeks she ran home and went straight to the bees, which she could see out her bedroom window. Bees can be left on their own for this long, but weekly hive inspections are the norm to make sure they are not getting too crowded (leading to swarming) or infected by mites or parasites (requiring treatment). Caroline was anxious and excited to make sure that they were OK in her absence: "I thought about them while I was on tour. One time I was driving in a car and a dead bee landed on my notebook. And I wondered if this was some kind of sign. I couldn't help but think about my bees and how they were doing." Even though she was physically separated from her bees, they still entered her imagination and thoughts. Caroline's concern about their well-being and ability to survive without her was a way for her to use the bees as an anchor to her own necessity on the planet.

Over and over again we heard similar stories of emotional attachments. Zack, a friend of Caroline's who is also part of an urban homesteading collective based in Brooklyn, spoke similarly about his relationship to his bees. While he does not "love" bees he has reverence for them and is very careful of them. When accompanying him on a rooftop hive check, we were prepped beforehand and told to be as quiet as possible. He told us, "I don't talk much while inspecting the hives. And I like to move fairly slowly." When the bees are producing he feels relieved. He likes to see them do well and gets excited when there are signs that they are healthy. While on Zack's rooftop everyone was impressed at the honey yield. Similarly, Eric conveys a sense of responsibility in beekeeping: "I find a sense of relief when we did a good job checking the bees and we have helped them. But when I am going in there I am disturbing them, there are things that I need to check, and this might cause things for them, they have to rearrange themselves. I am kind of a worrywart with everything. I am constantly worried about them. And I want to know how to fix things for them." Although Zack's concern is genuine, this points to the presumption that humans can know how to improve things for the bees. There is the hubris of human intervention, innovation, ingenuity, and know-how. When we nurture bees, sometimes there is a form of interspecies paternalism, in that "father" may know best. However, paternalism is part of a continuum of care, which can be distinguished from domination over natural processes via the cultural expertise of another/higher species.

Stewardship: The Responsibilities of Beekeeping

Eric Rochow, Zack, and other beekeepers who keep bees as a hobby rather than a primary source of income stressed that the rewards of beekeeping—in the case of an abundant honey yield—are not taken for granted. This practice emerges in what Jim Fischer called beekeeping "stewardship," performing particular duties and responsibilities, as well as an ethical component to monitor bees and aid them in their labors as best you can. In an interview at a hive on Randall's Island, Fischer explained it this way:

> The stewardship is like—if the bees are going to make me a little money through the honey and pollen, you know like stuff I can sell, then it is like the guy who has an elephant in Thailand. Elephants are their heavy equipment—they are the forklift, the front loader, the crane, everything. They use elephants for everything. The guy with an elephant makes his living off the elephant. So if times are tough, he will feed the elephant before he will feed himself. And he has to take care of the elephant in order to take care of himself. . . . That leads to the old statement about take care of the bees and the bees will take care of you. And that is true because if you take care of them they can make honey and all that.

There is almost a quid pro quo relationship implied within the relationality of bee stewardship. But in another sense, good bees can be viewed as the end product of good beekeepers/stewards. Honey conceived of as a relational object or process (the work of beekeeper and bee) has both emotional and economic currency.

Because Fischer talked about bees in a scientific and utilitarian way, we were interested in whether he ever had an emotional reaction to bees. He told us, "Of course, but I don't let that get in the way of being a good shepherd to my hives. I wait until I have done my work to react in any emotional way." He talked about grappling with the rationality of science amidst the awe-inspiring creativity of these insects:

> Yes, they are God's creatures, and that is my excuse for being intolerant of those who are less than diligent in their care and respect for them. But I keep bees precisely because I know that there is no God to protect them or anyone else. . . . I chose to protect my bees because I can, not because I feel I must. I don't like to see others approach it as a frivolous undertaking, as these are living creatures, and not mechanical toys. . . . In my view, one shows a deep and abiding respect for the bees (and all God's creatures) precisely by going to the trouble and care to set aside one's own emotional baggage, preconceptions, and metaphors, and thereby truly looking at and listening to the bees, or horses, or squash plants, or whatever. By trying to be as detached as possible, the beekeeper, like a doctor, can do a better job of not misinterpreting the behavior of the bees. As the decisions of the beekeeper can have a profound impact on the future of the hive, the beekeeper owes his or her bees a dispassionate and professional evaluation, and the best that science has to offer in the way of help when they need it.

This narrative reveals attachment in a more circuitous way. Fischer sets aside his emotion while working with bees but is highly emotionally invested in practicing a particularly rational and scientific type of beekeeping. In this regard, his feelings toward bees are transferred onto beekeepers, whom he views as being ineffective stewards. Specifically, he is suggesting that beekeepers put their human self (emotions and interests) outside of the practice of beekeeping in order to really be present for the insects themselves. Through a scientific, rational, and methodologically precise way, you can enter into a space of hearing the bees and have a dispassionate and yet compassionate relationship to them. Through this beekeeping methodology, he distances himself from other beekeepers he views as bringing too much of their own egos, their relentless human preoccupations, to the practice.

The disparate beekeepers in our study agreed that there is a strong emotional reaction and significance in finding dead bees in a hive. Eric Rochow emphasized being moved and also taking responsibility for the bees: "When I lose an entire hive it is really devastating. I feel deeply sorry. I don't know how you don't blame yourself a little: what could I have done differently?" Fischer, like Rochow, also implicated himself in the instances when he found that his bees did not survive: "I take it personally when they are dying because I myself have been stupid, I have been careless or not proactive. I have let something happen myself by not staying ahead of the bees." The act of "staying ahead of the bees" is a form of human/insect attachment. One in which humans do emotional labor on behalf of bees through a conscious and ongoing effort—anticipating what they might need, how or if to treat for mites, when to split the hives, and so on. We asked Rochow if he felt guilty when his bees died:

> Oh yeah. I was feeling really bad. Because when you pop the hives, they are head first in the comb because they are trying to get the last bit of food or they are freezing to death and they are trying to stay warm. It bums you out because you have hives, and you look at the frame and you got like thirty bees and their little rear ends sticking out and you know what you did to them. I am relying on these animals to help you and there is that sense of sentimentality of that. Clearly there is a cringe moment when they die. You do feel bad. It is the same as being invested in other animals. Animals like bees are more independent than other animals and they have lived for a long time without us. But at the same time, when something bad happens, it does feel like they depended on you because you put them in an artificial environment.

While Rochow expressed regret and sadness at his bees' death, and likened them to other animal/human attachments, he spoke of responsibility. He stressed that humans have controlled bees by acting on their behalf—containing them in a man-made and monitored hive. Beekeepers try and navigate this ambiguous line between helpfulness and harm through various types of stewardship. Yet how much is too much intervention? Implied within are warnings that we can only help out nature so much, because too much human intervention (however well meaning) can be injurious and destructive.

Perhaps the most emotionally descriptive of all the beekeepers we interviewed was Gita, who keeps bees in the Red Hook neighborhood of Brooklyn. When we asked her what she felt when her bees died, she explained how affected she was:

> I feel sad when they die. Our whole hive died and I didn't get to watch it because our hive blew over in the middle of the night. And we lost all the hive because they froze to death. And I was really upset, I actually didn't go back to the hive for like three months because I couldn't face it. Someone else closed it up for me because I couldn't deal. Because I felt like I had done a poor job making sure their safety was secure, I should have built a wind fence, I should have strapped them down. There was a whole bunch of stuff I should have done, and whether or not we did, it still could have happened but I wanted to have them around. I didn't want to keep starting over. I wanted to build on the hive. It is like a relationship and I didn't want to start over.

Her reaction to the death of her hive, possibly due to her own actions, underscores how seriously beekeepers take their responsibility for keeping bees alive. Her grief is comparable to what one would feel for a beloved pet—it was extended and highly emotional with a mourning period lasting three months.

Beekeepers become attached through stewardship and will grieve in the event that their bees die under their watch. Stewardship speaks to a particularly postmodern relationship to nature, what the sociologist Anders Blok refers to as a "risk-reflexivity" triggered by a growing sense that we have fatally harmed the planet so that the "survival and well being of animals emerges as a matter of human moral responsibility."[17] Beekeepers seem to stand in close witness to the inner workings of other habitats and networks and in the process develop an ethos of responsibility toward bees. Although bees don't really belong to humans, we feel compelled to help and manage them once we get more familiar, more intimate—for the sake of these bees and possibly due to a moral ecological imperative. As our responsibility grows, so does our emotionality, whether we are the self-appointed stewards of plants, animals, or insects.

Bees as Pets: Anthropomorphizing Insects

Most insects are not as popular as the bee—bees are on our cultural radar and we cross paths with them in everyday life. In general, insects of everyday life are to be avoided: usually they are often creepy, frightful, or unpredictable. Some species can potentially harm us—sting, bite, burrow, infect, destroy. When a bear accidentally wanders into a backyard or back road, we are startled at first because the animal feels out of place. We are cautious and mindful as we assess the situation and possible danger. But when a cockroach, beetle, or bee lurches or quickly flies into our apartment or house, our social instinct is to gasp, avoid harm, and flee. Often, we quickly decide how to get rid of the insect, the *bug*—through removal or death—rarely questioning the decision to do so.

Animals, whether bears or insects, fit into our codes and concerns, even our moral and ethical systems of belief. We classify animals under umbrella terms like "tame," "dangerous," "wild," "poisonous," "nocturnal," and "domesticated." Regardless of how bees are connected to our contemporary diets, ecological ideologies, and larger socioeconomic patterns of consumption, we do not typically frame bees as domesticated or tame, given their self-sufficient nature. Bees evade these targets of classification because they do not neatly or consistently fit into preordained human/animal categories. Interestingly, bees affect people like other traditional types of domesticated animals, those we call pets.

There is much evidence to support that people love their dogs—we name them, let them sleep in our beds, include them in family photos, buy them expensive cookies from dog bakeries, dress them in outfits that match our own, mourn them when they die, and bury them in pet cemeteries. Yet, as the political scientist Kennan Ferguson notes, dogs as pets "are in a servile position within the household. . . . There is clearly an imbalance of power inherent in pet ownership; that one party controls access to food, the timing of exercise, and the propriety of play (both temporally and spatially) bespeaks a clear domination."[18] However, bees, unlike dogs, are not as servile or dependent on humans—they do not wait for their "owners" to be fed or let free. Bees are self-sufficient and independent, and this translates into more complex and interesting relations of power between human and bee. They may feel like pets on an emotional and spiritual level, but ultimately they are not bound to us in the same way that we are bound to them. In this regard, bees have what the sociologist Jen Wrye calls "petness," defined here: "Petness which can generally be defined as the state, quality, or conditions under which a pet is constituted, arises from social relations and the treatment of objects. I contend that pets are a product of the investment of human emotion into objects, and that this is not exclusive to animals, but is also exhibited in our treatment of inanimate and inorganic entities."[19]

Petness allows us space in which to consider the boundaries that separate pets from nonpets. In this frame, anything can potentially be viewed as a pet by human beings, regardless of its constitution, whether it is a virtual pet, or for that matter a living snake, lizard, or bee—it's all contingent on the relationship the relevant human has with it.

Some urban beekeepers we interviewed think of bees as pets. One informant told us that his wife "sings to the bees" and another, Tim O'Neil, readily admits naming queens just as he would with any other pet. In response to the question "Are bees like a pet for you?" he quickly and enthusiastically responded, "Absolutely! They get named after I get to know them a bit. Oh, how do you name anything else? You get to know them, you give them a nickname. And then it just pops into your head—this hive is Matilda. This hive is a Lady Hamlet, this hive is a King Henry, constantly making new queens, ya know. You just get to know them."

O'Neil's description of how you "just get to know them" speaks to how people develop relationships with bees through different levels of interaction and attachment. Christine also admitted that in the past she has named her queens "Camilla" and "Beatrice," as well as the names of her husband's ex-girlfriends. While she didn't relate to her bees as pets, she said that they show a kind of intelligence. She agrees with other beekeepers who describe beehives as having identifiable "personalities," with some more active than others. Christine also mentioned how some beekeepers can hear bees' "moods" if you will, which depend on things such as the wind, cold, or rain. In this context, queens and hives are distinct from each other based on how humans perceive them acting to their environment over time, rather than to a particularly emotional exchange.

In comparison, Gita, who is a second-year Brooklyn beekeeper who tends multiple hives, said she has not named her queen like some people do. She explained her bees in juxtaposition to her pet cat and the maintenance of her plants: "I don't think of them like my cat who cuddles with me at night. I have a different relationship with them, a different ecology; it is kind of like plants. You have to nurture them the way you raise your garden vegetables and at the same time it is a little bit different because you have to watch their behavior a little bit more, although we do watch the behavior of the plants." Like others, she described the process of keeping bees as "an amazing thing to participate in and watch" as more of a "really cool phenomenon." But interestingly, she added that "you can have this really sincere deep interaction with bees," which seems to distinguish keeping bees from watering and observing plants, which are more static living objects, albeit a garden does involve a good deal of seasonal labor.

Beekeeping is indeed work, and people do hive checks to see if their bees are faring well and seemingly productive. We don't necessarily "check in" with traditional pets, because they are usually intertwined with our private spaces and part of daily routines. If they are mobile, responsive, tails wagging or purring contently, traditional pets have been implicitly checked. Checking in on bees is more of an event and ritual—you don't do it daily but it can sometimes be a hassle. Bees demand that you go to them, which can mean time, money, travel, and physical effort (particularly if the hives are on a rooftop three stories high). So unlike with traditional pets, there is a different type of site-specific and time-sensitive labor that takes place. This places an onus on the beekeeper to be present and even officious in a particular way. A responsible beekeeper knows what her bees are up to and can only verify this assessment through a visit.

However, the vast majority of our respondents scoffed at the idea of engaging with bees as pets, as if it breached some understood and accepted boundary—transgressing our cultural boundaries of animality. According to Eric Rochow, it's "silly" to name queens because it pushes "anthropomorphizing over the line." Bees have a relationship to other productive animals who straddle the line between exotic pet and utilitarian creature—most often they are discussed and raised alongside chickens and goats. People work with these species, nurture them in the same way as a cat or dog, but reflexively draw the line at some point—perhaps this has something to do with bees as a productive species. At times, you have to work harder on behalf of bees so that they make honey. There is the physical labor of setting up the hive, lugging the frames to rooftops, and climbing up stairs with hive tools and veils and smokers. For some, there is a fair amount of travel, whether an hour's train ride across town or a car commute upstate. The work is more ceremonial—it invites visitors and can become an event—particularly, as we found, in New York City where bees garner cachet for their human keepers. In a sense, the pet relationship is always corrupted or tainted by the desire to get the honey. Bees are cared for as pets but are also doing something that they do naturally under unnatural circumstances. When humans take bees honey, it is in no way a reciprocal act or collaborative relationship. A dog, as pet, can show signs of feeling loved (wagging, nuzzling, licking, draping its paws on its owner), but bees don't signal affection with such obvious gestures.

Rochow doesn't understand his bees as pets per se: "it is attributing a personality to something that it doesn't have." For him, bees are natural "machines." He told us that "they are not like dogs, the equivalent of two-year-olds. With my dogs I can tell them to do something, and they do it. When they are sick and they want to tell me to go out they can communicate." When we asked him about his comparison of dogs to bees he elaborated on the ways that bees don't "think":

> Their brains, their feelings aren't hurt if you don't friend them on Facebook. They have limited brain capacity and they are programmed to do certain things—that part is amazing. I roll my eyes at people who want to name their queen and yet I want to wrap it up and make it all warm for them. So that is the part where it's freezing out and I want to make it warm for them. That part is the anthropomorphizing part, where I want to keep them warm, and in reality it is OK for them to get cold.

He wrestles with his role as the bees' steward, which is a different role than being a pet owner, and he admits to worrying about whether the bees are warm and comfortable. This wrestling is an example of how bees are conceptualized as an independent species, a classification, but at the same time he worries over *his* bees. The relationality shifts, and his bees blur between "machine" and pet. It is not as easy to affectively disentangle the relationship—it seems as if the informants are struggling with this very issue themselves. How to both see the bee as this independent machine like a creature that works, and at the same time, as this vulnerable, fuzzy creature with a life span and a certain death—just like other animals.

The act of anthropomorphizing bees was brought up by other beekeepers we spoke with, mainly acknowledged as unavoidable and inevitable. Beekeepers engage in a type of negotiation between their intellect and emotion, as described by Jim Fischer: "People are allowed to think of the bees as pets. Some people sing to the bees. That is OK, it is kind of silly but it is OK. I preach against it because I don't like people to fall into the whole anthropomorphic trap. But it is almost impossible not to. It is impossible not to think of the queen bee as *her* as opposed to *it*. Or describe the workers as my girls or something like that. It is a mental shorthand."

We asked Fischer if he ever named the queen and while he said that he didn't, he offered that his wife "anthropomorphizes excessively." Hesitating, he qualified this statement: "Look . . . those things are actually good in that they promote paying more attention to the bees. Anything that makes you pay more attention to the bees is a good thing no matter the mythologies you tell yourself."

Bees can be described as pets, hobbies, and cottage industries—but never as companions or family members. There is no individual recognition for a particular bee, other than the queen and she is essentially in prison within the hive because she can never leave its confines (described in more detail in chapter 5). The queen's destiny is to produce for the sake of the colony, and her position (as a powerful reproductive captive) renders her at a further distance from humans. There is some ambivalence about where to put bees in the animal taxonomy, because we can't have a wholly physically intimate relationship with them. Nor can we truly invite them into our homes where they can fly around in all our nooks and crannies, be with us on our turf. We don't take them to the vet, tote them in pet carriers to travel across the country, or dress them in costumes. But we do use human terms, concepts, and distinctions to describe bees, and to some degree this anthropomorphizing brings the bee into our comfort zone. Even if we cannot engage with bees as part of our everyday lives, we have a means to approach them as a potentially knowable species and see them as more familiar members of common worlds.

The Sting: Insect/Human Penetrations

The instant you recognize that you are stung by a bee is startling. There are a few scant moments, seconds, in which your body communicates to your brain that some tiny breach has occurred. But, typically, before you can intervene, fluid from the insect's body is commingling with your own. It's an accidental chemistry that has tangible consequences for both actors. Usually it is an insignificant altercation from the perspective of humans (most people recover and live without consequence) but it marks the end of the life of the bee.

The sting is a site of vulnerability, a space or moment of significant interspecies exchange. A bee sting is different than a mosquito bite; insects breach our body's borders for different purposes. Female mosquitoes seek human blood for its nutrients to increase egg production, but bees do not naturally seek us out as a species to increase their survival. Humans are not food for the bee, unlike other parasitic creatures that feed off of us. While mosquitoes can transmit malaria and West Nile virus, for most Americans a mosquito bite means an annoying short-lived itch and site-specific swelling. Bees deliver more acute and lasting pain and, importantly, their sting is a defensive mechanism. Regardless, bee stings can be intense, long-lasting, and psychologically traumatizing. Depending on the area of the body, a sting can leave the site sore and swollen for days. Certain areas, like the face and hands, are especially vulnerable to the venom. Lisa Jean reflects on being affected, emotionally and physically, by being stung while in the field:

> *I am stung and I am instantly furious. My first instinct is to swat to stop the stinging or possibly to get revenge. I feel it instantly on my thumb; I've been penetrated. This piercing into my skin—the initial injection feels as if it is "not that bad" but quickly starts to ramp up as the toxins spread into my system. Disoriented, I step back from the hive—as I have been warned, once stung the bee releases a different pheromone, like a warning signal to the hive and other bees will be drawn to sting. I am stung in my lip and then in my neck. At first I am mostly embarrassed and don't want to let the other*

beekeepers know that I have been stung. Somehow marked as a novice or a fraud not really belonging in their in-group. I try to shake it off but each movement just increases the intensity of pain.

Trying to just ignore the sensation and the feeling is not possible and I am brought back into the sense of being a child, quite possibly the last time I was stung—when my interactions with insects and the "wild" were less stilted. The childishness of being vulnerable to insects embarrasses me further, as I yearn for a parental figure to care for me and make me feel safe again.

The neck sting is most painful at first but starts to subside quickly and I wonder if it was really a bee after all. My lip is swelling up and I can tell my speech will be impaired. But it is my thumb that is the most uncomfortable. It was precisely my novice approach to grabbing the frame quickly and without checking for bees underneath my finger placements that led the bee to sting me. She might not even have had a choice as I just placed my giant thumb on top of her.

It feels as if all my blood is rushing to my thumb and it is prevented from circulating back to the rest of my body. All of my bodily fluids are stuck in my thumb and it is aching in a rhythmic pulse. I keep expecting to look down at my thumb and see it has engorged to the size of an eggplant, but it is just slightly visibly swollen—almost disappointing me as I want more visual evidence so the pain is justified.

Not subsiding over time, the pain actually increasing. The lack of mobility in my thumb becomes obvious, as I can't bend at the joint. I keep thinking if I just put it out of my mind, the pain will go away and I will be able to focus again on the bees and the people around me. It is difficult to conceal the pain particularly as I am swelling up. As others begin to know I have been stung, they nod and tell me how it does hurt but it goes away and encourage me to move on and forget about it. But the pain goes on for about two days.

The body remembers the pain. The sting inflicts a sharp searing sensation, but also a mental and emotional one. A bee sting leaves not only venom and the stinger behind, but also a type of embodied knowledge—a memory. The immediate sharpness of the first few seconds of the sting is acute and searing: the body signals the brain that epidermal borders have been invaded. There is a breach, a shock, then for many an itchiness, heat, swelling, tenderness. With one exception, we have not met anyone who has been stung that *wants* to experience this accidental exchange again. (There are some who purposefully use the sting medicinally as discussed in detail in chapter 7.) On a very simple level, it's hard to imagine how such a comparatively small and unassuming insect can physically affect us to such a degree. The sting is real, but even those who have never been stung are afraid of bees. Why do even docile, calm bees send us violently swatting or ungracefully dodging at first sight of them even when we have no embodied experience to base those fears on?

A bee sting, a microscopic injection of fluids, can cause various degrees of swelling, burning, and numbness. A small percentage of the human population can become grievously ill, even die from the unplanned and unfortunate interaction. As B.J. recounts, even experienced people cannot escape being stung from time to time: "I swelled up from here to here like a sausage because I am pretty allergic. Most people are, but to varying degrees. But what most people don't understand is that less than 1 percent of the population is anaphylactically allergic."

Many beekeepers we spoke to see the occasional sting as part of the job and get stung in different amounts, with different levels of discomfort. However, Tim O'Neil, a beekeeper who is highly allergic, has to carry medication in the event he gets stung. He told us:

> Everybody poofs up. That tiny percentage of people that have to go to the hospital actually is me. I developed an allergy last year, but got it treated before it got to be bad but it is still a bit more risky for me than most people. I don't know if you notice but I carry an epi pen in my pocket. In the pocket of my hive suit. I think every beekeeper should carry an epi pen on them. It doesn't hurt and in the off chance that you ever need it.

An epi pen is short for epinephrine auto-injector, which is a compact medical device used to deliver a dose of epinephrine (also known as adrenaline) for those who may have acute allergic reactions. It is used to avoid or treat the onset of anaphylactic shock. Even though a sting can send O'Neil to the hospital, he is pragmatic, prepared, and undeterred. The work he does with the bees is worth the potential risk. He regularly uses a hive suit, veil, and gloves—full protective gear—when he gets close to his colony.

One significant point that links all beekeepers we interviewed was that they agreed that getting stung (or not) relies heavily on how you approach the bees and anticipate what they might be "feeling" that day, based on how noisy and active they are. They are definitely sensitive to rain and wind, as well as temperature. There are obviously some aspects of avoiding stings that are out of human control. In particular, abrupt changes in weather conditions can throw a serene visit into a chaotic and painful incident as Meg Paska describes tending bees in the rain:

> I am cautious and pretty methodical. The only gear I wear is a hat and veil. I wear normal clothes. They will get tangled in my hair if I don't wear something. I have actually taken a bunch of stings to the face before because I had to hive some packages in the rain and at the tail end of the day, the rain subsided and this last package was pissed and I went to shake them and they went right out of the whole and whoop, right up into my veil, and I got stung in the neck three times, and two in the head and I looked horrible. If you have short hair or no hair, you might not need a veil. But if you are talking a lot, then they go for the face. So I don't want to get stung in the face or the neck because that is the most dangerous place if I am going to have a reaction. I have had systemic reaction, no anaphylaxis, but my entire face breaks out and eyes get itchy. That freaked me out, but then afterward I got stung and it swelled for a couple hours and then it was gone.

Sara, a Bronx beekeeper, also told us a story about how precarious beekeeping can be, based on the environment and changes in weather:

> And a storm just whipped in and we were all wearing full bee suits and all of a sudden there were like a parade of bees and they were going directly for us and for this tiny little opening in our suits and for our ankles. So that we got stung a number of times that day and it was a really rough day and one person actually left the project after that day.

This experience, while physically dangerous, affected how she thought about her relationship to bees on a more philosophical level in terms of her role as a keeper and protector of a nonhuman species. It made her think about what she does for bees exactly, and how much control she has:

> It was so telling, that you actually as a human do not control what happens in a hive. All that you can do is make sure that they have enough space to expand and that they are reasonably comfortable in their quarters. I think that is when the hobby really changed for me and when I realized, this is me just making things available to me. They don't have to be there. They could leave at any time.

These moments, when bees in effect behave unpredictably and, in human terms, "freak out," turning on their keepers, point to the power bees can hold over us. And whether or not our work with and for bees is as necessary and beneficial as we believe it to be. Notwithstanding, beekeepers weather the sting again and again—it's a part of the job that can be managed but never eliminated.

Beekeepers learn to deal with the possibility of getting stung, but it is in no way a deterrent. In fact, we talked to people that appreciated, perhaps even anticipated, the likelihood of a sting. Negotiating the fear of being stung illustrates how working with the bees is affective labor. Affect, as the sociologist Patricia Clough has described, is "a preconscious, preindividual capacity, a bodily capacity, which, however, is expected to become conscious when put to language or when narrated, for example, as emotion."[20] In other words, the knowledge of a potential sting while not entirely conscious to each beekeeper (or even researcher) is lodged inside the body and memory, almost anticipated. This preconscious capacity to being stung is then managed, usually without being uttered. It is another form of labor that the beekeeper performs. Affective labor, broadly understood, refers to types of unseen and unacknowledged work that can unconsciously impact people. In a digital post-Fordist economy, the constant worry of monitoring and responding to emails not only at work but within the home is a form of affective labor that can make a seemingly benign task terribly stressful. Beekeeping also involves humans on affective levels. For example, beekeeping is affectively different than caring for dangerous animals such as a lion or bear. Both are potentially injurious for the beekeeper or zookeeper, but caring for bees involves more subtle affective labor. Yet it is more likely that you will get stung than bitten. Affect matters here because it reaches the beekeeper on different planes; it's not simply pragmatic.

As Paska explains, getting stung by bees can be emotionally positive. It's a human/insect experience in which, to a certain degree, the bees are in charge. She explained the feeling of being able to conquer her fears and the potential of the pain of the sting to us candidly one summer afternoon in her backyard in Greenpoint, Brooklyn (while watching some of her rooftop bees dip into her neighbors' plastic kiddie pool):

> It sounds a little silly, something about handling bees is tremendous for the self-esteem too. I mean I am always like, man, I am such a fucking badass [smiling but sincerely]. I am holding bees, these frames of bees and they could just sting me if they wanted to and they don't. It is awesome and it is kind of, like, I don't like to pat myself on the back, and I want to congratulate myself a little bit.

Without a doubt, there is a daredevil aspect in working with thousands of unpredictable insects, no matter how knowledgeable and attuned to them you may be. For us, as novice researchers, we both understand the feeling of being "a fucking badass" for getting close to bees, relinquishing fear and control throughout our fieldwork. Our own self-esteem increased every time someone told us, "I could never do that!" We now understand how working with bees could be invigorating, as well as how it could make you feel special, in some way "cooler" than others.

We heard countless sting stories throughout our time in the field, perhaps subconsciously activating our own fears of being stung while doing hive inspections, or just simply talking with people a few feet away from an active hive in which bees are unmistakably rotating and spiraling the space around us. It was interesting to watch some of our informants casually walk around and sit nearby their hives, while they were clearly used to being with their bees; sometimes, particularly when we were new to the field, we had to work to mask our anxieties. Part of beekeeping, for those informants with more accessible backyard hives, is simply being around the bees, which is different than actually opening up the hive to inspect, check, and potentially move bees around as part of normal spring and summer hive maintenance. As researchers, the bees were always confidently, perhaps even boldly present, sometimes on occasion drowning out our human subjects' explanation and stories of insect/human exchange. When a bee enters your personal space you can do two things: stay still and let the bee walk or rest on your body where it wants, or try and move, swing, or shield yourself so that you avoid it. Whether you ignore (remain still and calm) or move to avoid the bee, the insect has caused you to react. When bees move or fly toward humans, we respond and act, but it is the insects that are acting upon us. We have learned to heed and watch for flying bees, because on some level most humans (laypeople and beekeepers alike) are afraid of getting stung.

But there is an implicit machoness related to the sting; the unspeakable toughness or "badassness" sets you apart from other cautious and protected beekeepers. For some, it's as if it is supposed to hurt if you do it the right way—the *wild* way. Maybe the bee sting makes people feel something more. The world disappears around you and you are in your body, in the moment, full of endorphins, excitement, and thrill. There is also that *being there*ness of working with and interacting with the bees without protection. You get a momentary out-of-body experience through being penetrated by another species. This tiny insect moves you to consider, however briefly, the breach it has made and the pain it has caused. Nothing else exists and there is pleasure in focusing on the pain. This type of pain is caused "innocently" and "naturally," and we take it because we deserve it and it feels good. This is comparable to what the sociologists Patricia and Peter Adler describe in their book *The Tender Cut*, which explores self-cutting among teenagers.[21] Rather than reducing self-cutting to a sign of suicidal behavior, the Adlers argue that cutting can at times be therapeutic because it helps deal with emotional pain. At a public lecture Comfort began by taking off his flip-flops and said, "This is also how I tend bees—barefoot. The bee sting is my favorite part of keeping bees. If you don't get stung than the honey isn't as sweet. I like getting stung. I never wear any gear."

Interestingly, two of our informants, Sam Comfort and Jim Fischer, share the same philosophy about communing and commingling with bees without any protective gear. For Comfort it is more natural and wild, and for Fischer it is better for the bees—or perhaps a sign that he tends bees the right way. He told us, "When I work with bees, people don't walk away with stings on them. I am fairly slow and cautious. I encourage people to work with as little protective covering as possible. And no iPod. Listen to the bees: you are there to enjoy them. They do not know they will die when they sting you, but they warn you." Using a more controlled approach effectively protects you from insects that are out of control. The prophylactic is not in specific gear but in the deeply empathic and somatically engaged practice of keeping attentive to bees. In this way, listening to bees can protect you.

What we have been working toward in this chapter is a way to describe the layers of entanglements that materialize between bees and humans. From a bee getting stuck in human hair and buzzing closer and closer to an ear to bees dying in boxes on apartment building roofs after a brutally cold winter season, humans and bees intersect through emotional and physical entanglements. The intersecting nodes of insect/human are emergent; in these shifting spaces, bees affect humans both productively and painfully. Exchanges between beekeeper and bee reveal the significance of the buzz, the smell, and the sting as forms of cultural exchange and an intimate emotional and spiritual exchange between species. Bee love is a complicated ball of wax, whereby there are real feelings and emotion expressed at the awe and wonder of insects. And at the same time there is a sense of otherness and serious confounding emotions that make it seemingly impossible to break through to interspecies flows. But the sound of the buzz, the taste of the honey, the smell of the hive, and the fear of the sting speak to how bees are connected to unregulated sensibilities—affective facts that exist outside of experience. Bees are not simply insects or bugs: they are vibrant matter. Enmeshed with bees through body, mind, and soul, urban beekeepers celebrate and fear the vibrant matter that is co-created in their mutual practices.

5

Entangling with Bees

Sex and Gender

As you might expect, the walls of Sunflower Academy in Prospect Heights, Brooklyn, are covered with letters, numbers, shapes, and colors arranged in a slightly frenzied yet organized manner. Once you've entered and turned to face the cubbyholes, you are met by smiling, cheerful yellow and black bees that adorn the walls, with the date and days of the week written on their abdomens. On this particular spring day, Lisa Jean's eighteen-month-old daughter's teacher had organized a Bee Day that featured special guest visitors: seventysomething beekeeper Farmer John, an affable man dressed in baggy overalls and a straw hat, and the real stars of the show—a box of live bees safely contained behind glass. Twenty-five of Brooklyn's tiniest hipsters, from one to three years old, crowded around Farmer John's observation hive, peering in at the insects and excitedly pointing to show their teachers (see figure 5.1). Children were encouraged to handle many beekeeping tools, smokers, bee suits, veils, and brushes. Plush bee stuffed animals and shiny plastic bee models were arranged on the tool table. Then sitting "crisscross applesauce" on the colorful patterned rug, the children, at varying degrees of attention, listened to Farmer John talk about beekeeping. With the help of a bee hand puppet and some large posters of bees performing various tasks, Farmer John told the "boys and girls" how very important bees were to all the food they eat. The next generation of urban beekeepers was on the way to caring for their own rooftop hives.

Pausing dramatically, Farmer John offered, "The queen bee is the mother to all the bees—the mommy. The drone bee is the daddy of all bees. His only job is to fly up in the sky and meet the queen and then they get married. After he gets married, he dies. The queen then comes back home and has all the babies you see in the hive. Every bee you see is the baby of the queen mommy. That's a lot of children she takes care of." Clearly the adults were the only ones startled by this

Figure 5.1. The Sunflower Academy checking the observation hive. (Photo credit: Lisa Jean Moore)

description—we exchanged looks with one another, mouthing "dies" and shrugging. But scanning the audience, the children were unfazed and continued to watch Farmer John as he ambled up and down the makeshift stage in his out-of-context country attire. He then showed the children some poster-sized photographs of queen bees and drones. A safety discussion of bee stings followed. "Now, bees, they can sting. When you get stung, you have to run away, tell a grownup, and get the stinger out with the help of a grownup and put ice on the sting." Some of the daycare workers were sharing sting stories in hushed tones. "I really hate bees," the woman next to me shivered.

Farmer John also brought some props to teach the children about pollination. Every other child was given a silk flower to hold. They were instructed to stand up. The remaining seated children were fitted with antennae headbands. The children were then told to walk or skip around the room and those with antennae were shown how to

Figure 5.2. Buzzing the room in pretend bee play. (Photo credit: Lisa Jean Moore)

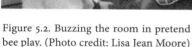

Figure 5.3. Farmer John gives honey tastes. (Photo credit: Lisa Jean Moore)

put their faces into the flowers and say "BUZZ" really loudly. Instantly the room was alive with toddlers, chanting *buzz* and banging into one another as they shoved flowers in each other's faces (see figure 5.2). A combination of bewilderment and frenzied energy led to several children crashing into one another and some children falling on the carpet, laughing and rolling about. Clearly these children were not able to perform pollination with the same choreography that the bees used. The finale of the presentation was tasting honey from Farmer John's bees. Children lined up with their index fingers extended. From a plastic honey bear container, Farmer John squirted a bit of honey on each finger. Several children returned to the end of the line to get more honey, clearly enjoying the sweetness on their sticky fingers (see figure 5.3).

In this classroom visit, these preschoolers caught the buzz. As Farmer John packed up his traveling hive, their teachers attempted to settle them down. This was a normally difficult task made even more challenging, as their bellies were full of honey and their little bodies bounced off one another as many pretended to still be bees. In keeping with the theme of experiential learning, for even the littlest student, the art teacher cut up bee body parts for a pasting and painting post-nap activity. Even in Brooklyn, a place seemingly disconnected from the field and farm, bees are now becoming integrated into the curriculum. This pedagogical use of the bee is part of the larger beekeeping buzz in New York City, as well as a recent interest in teaching children about sustainability, local food, and being more "green." A group of urban preschoolers was initiated into the world of the bees and how they are part of our food production and environment, but they were also taught another lesson about the world of adults and "mommies and daddies"—one about gender, sex, and productivity.

Birds Do It, Bees Do It: Sexuality and Honeybees

All bees, in particular honeybees, are anthropomorphized in a way that many insects and most animals are not (save for the domestic dog or house cat). Bees are described as industrious, helpful, driven, purposeful, cooperative, and smart. Often illustrated as "cute" in appearance, their striped furry bodies, wings, and round eyes have long made them ideal subjects for children's clothing, toys, and books (an Amazon.com search of the words "children's books and bees" yielded 2,064 titles). The colloquial phrase "teaching about the birds and the bees" is a convenient and well-understood euphemism that describes the intersection between the cultured socialized self and the "natural" desires of the body—or more simply put, we use bees (and other species) to talk around and about human sex and sexual reproduction. Obviously it is a tremendous leap for anyone—adult and child alike—to get from bees "getting married" in the sky to sexual reproduction, to familial relationships, but using other species has been a staple of socializing children to the knowledge of "how they were born."[1] But it is not only biological processes that are narrated for children: cultural expectations are also tacitly transmitted in the ways we tell them "the facts of life." The ubiquitous presences of heterosexuality—that it leads to marriage—and gender—that boys are daddies and girls are mommies—are often depicted in how bees, among other creatures, are shown participating in conception. These stories help to cement the naturalness of heterosexuality throughout the animal kingdom. What is fascinating about Farmer John is that he in effect explained social reproduction through what we sociologists would call "heteronormative master narratives" of insect/biological reproduction. Ironically in this case, the bees are used to show how the animal kingdom asserts the naturalness and superiority of heterosexuality. Yet these ideas about heterosexuality are what frame our understanding of mating in the first place.

Toddlers are actually being taught multiple messages through the medium of the bee as a vehicle of communication. Humans have long inscribed bees with meanings to broach the subject of burgeoning sexuality, as well as gender relations and even the importance of a work ethic. To be as "busy as a bee" speaks to how we value industriousness over indolence. The term "queen bee" is synonymous with the "it girl," the one whom we are socialized to aspire to be in contemporary American popular culture. In both instances, she occupies a particular social and hierarchical location that signifies her uniqueness, draw, and power. In this next section we examine how bees, arguably the most sexed and gendered members of the insect world, are spoken about by urban beekeepers. The projection and transference of gendered ideas onto the queen's reproductive body, the worker bees' domestic maintenance, and the drone's single focused insemination flight lead to vibrant descriptions of highly sexed practices.

The queen is the singular reproductive powerhouse of the hive or the colony, and her status within the hive is discussed with a fascinating combination of awe and admiration and resentment and sexism. She is arguably the most important member of the hive because a hive cannot exist without her. When bees swarm, the queen leaves the hive surrounded by a group of worker bees to establish a new colony. There can be thousands of workers in this migration, but there is typically one queen. Even though she is always the numerical minority, all activity revolves around her well-being and ability to produce eggs. The workers keep the queen safe, fed, and cleaned as she focuses on her one crucial task. In addition to holding so much power in the colony, she is the most humanized and metaphorically illuminating insect with regards to gender politics.

But these "facts" of the queen's life span are often conveyed with deeply gendered and socially laden descriptions. For example, we have heard the queen referred to as a "bitch," "promiscuous," and as a "loose woman," and comparisons between the queen and "typical" female behavior are common. In a lecture at a beekeeping meetup class, our instructor, Jim Fischer, explained the sexing of bees in terms of genetic heritage, stating that "all drones are bastards" and that "relations really matter for bees as bees seem to know who is their full sister or full brother." He further explained that "if you are a drone, then you have no father." Ultimately, we were told that "genetic loyalty is always to the queen." In popular representations of bees, as well as within the specialized world of beekeeping, the gendering of bees is related to common Western beliefs that position women as more natural than men, in large part due to their reproductive capabilities. The queen, as well as bee "family life" and "sexuality," reflects cultural assumptions of how females are closer to nature than males (those who transform nature, i.e., "man-made") and how nature itself is discussed as female (i.e., Mother Earth).

Through anthropomorphism, beekeepers and citizen scientists alike narrate the lives of bees as exotic and familiar. In order to make bees familiar, we rely on existing stereotypes like "the battle of the sexes" or "policing the borders."[2] When bees are benign they are cast as behaving more like "us" and when bees are threatening, they are cast as "others." Gender is one way to better understand the species in familiar terms that are meaningful to us.

The slippage between human gendered narrations of wholly insect behavior happens simultaneously with the actual changing gendered human world of urban beekeepers. So just as humans are narrating bee behavior in ways that are intelligible to humans, beekeepers also narrate human behavior in similarly gendered ways (see, for example, figure 5.4). When discussing the fear of bees as a problem for legalizing urban beekeeping, Fischer stated, "Overprotective mothers overreact to bee stings. Suburbia is no longer bee friendly; there are exterminators to get rid of bees. Allergies to bees are overexaggerated. Overprotective mommy thing. I mean, come on, there were no helmets, no seatbelts when we were young." This statement implies that it is women who may be obstacles to beekeeping being seen as a legitimate practice in the city.

Figure 5.4. Cover art for a children's book. (With permission from M. Maitland DeLand)

Surveying the room and noticing that the majority of seats were filled by women, Fischer paused and considered himself.

Fischer then switched to discussing how the demographics of beekeeping are changing. Perhaps realizing he had overstated women's negative role in the bee's bad reputation for stinging, he mentioned that now women were starting to join beekeeping in droves: "For the homeschooling mom, beekeeping is a great science experiment. And the amount of testosterone needed to keep bees has gone down. So now there are more women than men. It's not as macho anymore. Beekeeping was a brotherhood and now it is a sisterhood too." The introduction of more women into the urban beekeeping community could indeed modify the ways in which people view bees as dangerous to handle and be close to (requiring "testosterone"). Women beekeepers may be seen as evidence that bees are more "safe" than previously thought. An increase in women beekeepers may also influence how the queen is characterized.

The Queen

A noted Cornell entomologist and guest lecturer to the meetup group, Thomas Seeley, introduced the queen this way: "She is the anchor. She is the genetic heart of the colony but she does not get involved in decision making." There is typically only one queen per hive. However, new queens are introduced for a number of reasons; the current queen is old or ill, the hive is too small and will swarm, and/or the workers determine the queen is not laying enough brood. Brood laying is intensely monitored by her attendants who literally follow her every move around the hive. Our own observation of the queen in several hives led us to frequently comment about her lack of autonomy: she is hassled at every turn and crowded in her seemingly intimate work of laying babies. Her mother, also a queen, will place an egg in a special cell called a queen cup. This larva is raised on royal jelly—a secretion from glands on the heads of young bees, and most commonly the old queen will leave the colony prior to the emergence of a new queen. Once "supersedure," or the process of replacing the queen, is completed, the new queen, called a virgin queen, will find and sting to kill all other virgin queens, establishing herself as the only queen. Queen bees can sting repeatedly without dying, unlike other bees in her hive that sacrifice their lives after stinging. The virgin queen flies out of the hive to an area of drones and mates in flight for several days. Once she is fully mated with upward of a dozen drones, the queen will amass the sperm (roughly 5.5 million)[3] in her "spermatheca," or the receptacle in the reproductive tract of invertebrates where sperm is stored. The queen holds the sperm and releases it throughout her life span as she lays eggs. Queens live up to between two and seven years and at their peak they can lay roughly 2,000 eggs per day. Over their lifetime, they could lay a million eggs.

Attempting to describe the birth and death of bee queens without any metaphorical flair is challenging. Perhaps the best approach is to understand that all scientific discovery, particularly when investigating animals, is made comprehensible through the use of metaphor—and while these metaphors are powerful tools that enable us to relate to the "natural kingdom," they also limit what we might be able to see. The life course of the queen lends itself very well to embellishment and anthropomorphic description. Take, for example, the beekeeper William Longgood's description of a hive's new queen:

> In literary terms we have here the ingredients for a first-class Shakespearean drama: The reigning queen witlessly bringing about her own ruination by providing her successor, a secret privy council plotting against the throne and selecting from among candidates in the nursery the one to become the new regent, keeping the reigning sovereign in the dark so she will continue to serve until convenient to oust her in a coup d'état, palace intrigues among the conspirators, the double bloody deeds of regicide and matricide by an assassin chosen by lot. We may further garnish our fanciful plot with a bit of sexual titillation: the handsome young prince who will win the beautiful princess but only at the cost of his own life.[4]

Evoking Elizabethan-era plots, intrigues, and conspiracies, the internal political management of the hive centers around the queen's well-being; however, she is kept as a prisoner. The use of the word "queen" to describe this bee (and drone as well) originates from Charles Butler, Queen Elizabeth I's royal beekeeper in 1609, when he wrote "the Q[u]eene-bee is a Bee of a comely and stately shape."[5] Each urban beekeeper is quick to point out the pedigree of a queen, the place she was purchased, her cost (anywhere between $20 to $150), and how many seasons she has lasted. They describe their queens as "good layers," "fat," "quick," "nimble," or "duds," "lazy," and "sickly," but they are also described as "sexy." In this passage from the book *Plan Bee* (2009), the beekeeper and nature writer Susan Brackney likens sighting her queen to spotting an elusive celebrity:

> I often try and spot her on the job, but that isn't easy. When I examine her lower reaches of the hive to see just how many baby bees are produced, the queen runs like a starlet dodging the paparazzi. . . . Occasionally, I've glimpsed her running away, her attendants in tow, and each time is exhilarating. I feel as if I have seen someone famous. Or at least, someone really important. My sexy Italian queen is slender . . . her caramel-colored abdomen brims with all the eggs and sperm she'll ever need.[6]

Here the queen is narrated as an exotic beauty, both racialized and sexualized, albeit reproductively autonomous. In this account, glamour and mystique transform her into an insect beauty queen, as "she's the one wearing the tiny gleaming crown."

As a means of establishing a relationship with these queens, some beekeepers name queens after ex-girlfriends or famous women in history. Two clever beekeepers who wanted to get an endorsement from their synagogue for use of the rooftop had a contest to name the queen bee of the hives. The winning names were Eve and Rachel. But there are some who avoid the practice of naming bees, perhaps because of the shame of other people's presumptive practices, as Meg Paska states:

> I don't name my queens. I am not a big fan of naming queens. I mean that I understand why people do, but for me I hold my bees in such high regard that the idea of naming my queen just sort of strips her of her dignity. I don't want to do it. Bees don't give each other names and I am not going to impose that on them. I would rather them just be what they are. I don't like to think of myself as a bee master lording over them. I like to think I am helping them out and they are helping me out and it is a mutual arrangement. I take care of them and try to make the conditions right for them to live.

Here Paska is truly adopting the companion species relationship. She is relating to the bees in a way that considers her bees' existence as inescapably entwined with her own interests. As Donna Haraway, specifically writing about dogs, has explained, our companion species connections are "kinship claims" whereby the term "species" is about defining difference and enabling the possibility of "relating in significant otherness."[7] Haraway understands that when we have dogs in our lives it means we must have a relationship where humans lay out strict regimes of discipline and order (leashes, identification, commands) as a way to enable dogs to flourish in urban environments that are dominated by humans. So instead of seeing dogs, or in this case bees, as equals in the relationships, Haraway instructs us to understand that when animals are inhabiting urban habitats, humans must take on the responsibility of creating conditions that allow them to thrive. Companion species relationships are possible between all animals, nonhuman animals, and humans. As Paska explains, although there are "mutual arrangements" where animals of different "species" can form relationships of intimacy and affection, she is still in charge of taking care of her bees.

But despite the attempt to foster intimacy with the queen or respect her as a dignified member of your hive, queen bees do have a "sexual reputation." For example, during a lecture about reproductive practices of hives, our instructor explained, "The queen is promiscuous and wants to mate with twelve drones. And being a loose woman does have advantages for your children. Worker bees are neuter or neutral. They cannot lay a fertilized egg. It happens very rarely and egg policing takes care of it." Queen bees are the only ones who have both sex and gender in that they are sexually promiscuous and also not neutered. Worker bees are female bees but they do not have the reproductive capacities of the queen, nor do they mate through sexual intercourse. However, sometimes a hive will have a laying worker bee. In this case, a worker bee with functioning ovaries can lay unfertilized eggs, a form of parthenogenesis. These unfertilized eggs result in drone bees, the male bee. When a beekeeper opens up a hive and sees more drone brood than expected, indicated by larger, rounded capped honeycomb cells, he or she will often suspect that the hive is queenless. A queenless hive is disastrous in the eyes of a beekeeper because it may mean that the hive has swarmed or the queen has died, and the reproduction of the colony is then in jeopardy.

Accompanying Jim Fischer to visit the rooftop hives of the Five Borough Vehicle Depot's Green Roof on Randall's Island, Lisa Jean assisted with checking the boxes. As Fischer worked the hives, he narrated his efforts:

> I don't think this is getting ready to swarm, so I don't think they like this queen too much or what is going on. Did she die? I am looking for her. But she has a lot of brood. I don't see her majesty. She might be gone? I am trying to find her so I can see why they are making the queen cell. She must have ran on me. I am not going to bug her. I am going to assume I have a queen in here. All had larvae. Not as successful as that hive but still OK. Have to check it again. I am going to give it some protein to pump it up in case we get bad weather.

Because "queens always run for it at the slightest disturbance," according to Fischer, it takes a quick eye and agile handling of the frame to find queen bees. Typical of many checks for the queen, beekeepers go through each frame of a box to attempt to locate her. Beekeepers must assess if the queen is in good health and plump. As novice beekeepers, we experienced the excitement of locating a queen and being the first one to find her. Our cries of "there she is" were often met with accolades from more experienced beekeepers.

Drones

Crucially important to the health of a hive, queens are, however, not the only reproductive agent within a bee colony. Queens cannot do their jobs without drones. Popularized by Hillary Clinton, the phrase "it takes a village to raise a child," attributed to an African proverb, applies to the way hives are classified. This speaks to how humans characterize bees as particularly social insects.

Drones are the males of the honeybee colony. As providers of sperm, drones, unlike worker bees, have remarkably little else to do to contribute to the health and sustenance of the hive. Born from unfertilized eggs, drones do not have stingers. Christine, a Manhattan beekeeper, thinks the drones are "pretty much useless—just flying around, looking for sex." As a writer, Christine finds it fascinating to examine the etymology of the word "drone" and how it came to mean a lazy or slothful person. A drone is synonymous with a loafer, or a person who lives off the work of others.[8] She thinks that it is interesting that early beekeepers such as Virgil incorrectly believed that the hive had to have a king "because how could an entire species exist without a king?," implicitly questioning the "natural" patriarchal order of the animal kingdom. And it wasn't until the microscope was invented that scientists could identify queen bee ovaries "and this would mean that it was the female running the show." Notions of supposed male preeminence in reproduction are not limited to bees.[9] During the 17th century, some of the first scientific ideas about the form and function of human sperm were developed. The theory of "preformation" asserted that within each primordial organism resided a miniature, but fully developed, organism of the same species within a sperm cell. Women were considered "mere vessels" in the context of human reproduction, leading to what the historian G. J. Barker-Benfield calls the "spermatic economy," in part because semen was believed to embody men's vital life force.[10] Biology is cultured in both species.

Originally seen as a male-dominated species, drones seem to have been recast as problematic members of the hive. Perhaps this is because drones are useful only for mating and are not good "providers" and "fathers" as it were; they don't participate in daily hive life. They are crucial for the continuation of the species but they are akin to men who don't work hard or care for their families—in colloquial terms, they are just dicks. Drones don't do masculinity right within the context of the hive hierarchy, which is gendered as social and reciprocal—everyone is working hard continuously except the drone, who has one job to do and then dies. The degree of antipathy toward drones is evident through an analysis of our interview transcripts; drones are described as "useless," "lazy," "in the way," "stupid," "honey gorgers," "unwelcome," and "incompetent." One informant shared a particularly objectionable view of drones with us: "All drones do is eat, lie around and smell bad, and wait to mate." It was surprising to hear any one bee described as prone to lying around and so inert, given all of the larger narratives of bees as a buzzing, kinetic, productive species. Drones don't fit in with the rest of the group and are outsiders.

As described in more detail below, there is concern about the hybridization of bee colonies. And who is considered responsible for this hybridization? As Fischer taught us, "All drones are out clubbing and it's ladies night. It is like sending your daughter out. Who is she talking to, it's all about pedigree and diversity with feral drones. We as beekeepers are interested in where drones hang out, what are the changes in the landscape, like a river or a park, and does this influence where we can find the drones?" Jim is implying that queens, like daughters, inevitably need to be allowed to reproduce; however, they are also in need of protection to ensure the production of the right type of offspring. Drones can literally be bad seeds. Returning to Longgood's delightful literary analogies:

> The virgin queen is pursued not only by drones from her own hive, her brothers, but also by those from neighboring hives, all drawn by her

> scent and signals. Often the drones gather in common congregating areas, spending their days in idleness, like so many loafers in a neighborhood park or pool hall, perhaps boasting of their prowess, but certainly not of past conquests. They engage in daily reconnaissance flights, a kind of cruising, to sniff out receptive females.[11]

True to Farmer John's lesson to preschoolers, mating flights are death sentences for drones because when they insert their endophallus, it is ripped from their body during intercourse. The straightforward, arguably macabre, physiology of bee reproduction directly cements sex and death. There is a perverse violence in this act when considered under human norms regarding insemination, birth, and death. Maybe the lack of masculine domination is another reason why drones are so often pejoratively characterized. The role of drones doesn't fit in as neatly within the hive hierarchy or within human gender norms that advance the idea that masculinity is defined in part by physical labor. The queen is demarcated in large part by her reproductive labor; she is gendered as the ultimate mother figure. At the same time, the more problematic drone does not come even close to fulfilling the amount of labor associated with traditional notions of masculinity. The queen may be a "bitch" or "slut," but she proves her mettle though tangible evidence of her work—the brood.

Human occupational sex segregation, a socioeconomic concept, examines how historically women and men generally have not worked in the same occupations or industries. Typically women are administrative assistants, nurses, preschool teachers—occupations that are defined in part by caring and helping others through emotional labor. This type of work performed in the labor force mirrors that which women often do in the private sphere of the home without pay, such as caring for children, organizing schedules, and arranging doctor appointments. There is a gendered division of labor inside and outside of the home. On the other hand, men are firefighters, electricians, and politicians, positions that often require a form of emotional detachment because they are considered "official" public professions executed in the public sphere (rather than in semiprivate/public spaces such as classrooms or hospital rooms). Traditional male occupations such as these convey a degree of prestige and authority, and, in the case of politicians, economic and cultural power. Overall, men have a higher status and have higher-paid positions than women do. Additionally even when adjusting for occupation, women are compensated at lower rates than men. For example, the 2010 Joint Economic Committee of the United States Congress determined that a persistent wage gap exists between the sexes whereby women working full-time, year-round earn 77 cents for every dollar earned by men, and virtually no progress has been made in closing the gap since 2001.[12] In our species, women are typically in sex-segregated occupations that are less prestigious and less compensated. As beekeepers narrate, the hive, however, presents seemingly reorganized occupational and labor-status hierarchies.

During early spring hive checks, beekeepers are hoping to find frames full of brood. These capped cells indicate that a queen is laying eggs and that she is healthy, readying the hive for spring foraging and honey production. The drone brood, however, is not met with such enthusiasm. These round, dome-shaped cell caps are necessary in small numbers to maintain the reproduction of the hive, but beekeepers express anxiety when seeing what they consider to be an excess of drone brood. Too many drones mean more mouths to feed with limited labor power in return. During a hive check with Sam Comfort, he shared, "Oh, don't really want to see that. That's a lot of drone brood. So either I've got a laying worker, or a queen that hasn't mated properly and she is laying too many unfertilized eggs. Let's see if we can't find her. Yup, I think I might have to requeen this hive." Here the evidence of a high percentage of drone brood means that a queen has "failed" and must be "replaced."

Larvae are more vulnerable since they are more attractive to parasitic mites, like *Varroa*. Because the drone incubation period is longer, this allows the *Varroa* more time to become established within a hive and infest others. Beekeepers, leery of mites and displeased with drone brood, sometimes choose to uncap the drone brood to inspect for *Varroa* mites. The uncapped brood is then frozen for twenty-four hours to kill mites, then thawed for twenty-four hours before placed back into the hive. This conserves the comb, the structure of the hive where larvae incubate and honey is stored, and this enables the bees to reuse the now-cleaned cells. Drones are seen as potentially devastating to the colony as well as dispensable in times of *Varroa* inspection.

Our examination of drones would be incomplete without discussing another permutation in the meanings of the word "drone" in the 21st century. "Drone" has taken on a quite different meaning as the United States continues to engage in wars—most people now think of unmanned vehicles that are remotely piloted and used to drop bombs or practice surveillance. As explained by the literature professor Daniel Swift, contrary to conventional wisdom, these aircraft are not called drones because of the sound they make.[13] Rather, in the 1930s, the United States Air Force experimented with an unmanned aircraft that they called the "queen bee"; as this aircraft was modified, it came to be called a "drone" because a drone follows the queen bee.

The geographer Jake Kosek argues that the honeybee is a species often used metaphorically in the war on terror.[14] In particular, drones are typically used for tasks that are deemed too "dull, dirty, and dangerous" for humans. Military or predator drones make flights over battle zones in countries such as Afghanistan, Iraq, and Pakistan and have been praised by those in the military as necessary tactical weapons. In September 2011, military drones were infected with a "mysterious computer virus"[15] that captures the keystrokes of the pilots operating the unmanned aircraft. In both bee culture and military culture, the role of drones is reduced to the performance of a narrow set of "heroic" duties. Surveillance, bomb dropping, and insemination—all masculine activities—are connected to weaponry and, ultimately, death for bees because they die after they inseminate the queen. These drones are not considered flexible in the sense that they are "programmed" to do only certain things. Their work is specialized, specific, and covert.

Workers

Much to our surprise, eclipsing the queen, the worker bees garner the most praise from beekeepers. With frequent expressions of fondness and affection, beekeepers discuss their "girls," or the "ladies," as the most impressive members of the hive. One beekeeper even called her workers "feminist sisters." Worker bees are sterile female members of the colony and have a highly specialized division of labor among themselves and across an individual bee's life span; worker bees can be cell cleaners, nurse bees, wax producers, comb constructors, nectar processors, foragers, house bees, guards, and water carriers. Unlike the drones, they are flexible in that they perform many duties and they work nonstop. The majority of this work is comparable to traditional women's work within the private sphere, in particular, tasks like cleaning, cooking, and caretaking that are performed every day. For women, work such as this is often unpaid, unrewarded, and even invisible within the home, yet the labor of female worker bees is lauded and respected. Unlike drones, the workers are usually framed as integral to the hive community. Healthy hives contain up to eighty thousand worker bees that are able to keep the hive at a consistent temperature, regardless of the exterior temperature, by fanning their wings and clustering together. Because bees do not hibernate, it is crucial that worker bees be able to manage the hive throughout winter with honey stores, temperature regulation, and house cleaning. Beekeepers make note of the impressive skill with which worker bees launch from the landing pad and navigate their return full of pollen or nectar. The sociologist Arlie Hochschild's concept of the "second shift" suggests that demands for working women to both work in the paid labor force and also care for the family and the home create inevitable inequities in labor between the sexes.[16] Worker bees are in many ways valued by beekeepers, particularly women, for their industry and talent at "keeping it all together." This is another way in which bees are seen as "model insects," albeit in a very traditionally gendered capacity.

Indeed, as argued by the anthropologist Anna Tsing, North American beekeepers see an affiliation between honeybees and humans, particularly as the hive is "emblematic of domesticity."[17] Devoted to home and hearth, worker bees are sacrificial to the collective offspring, diligently gathering food, and tending to cleaning, protecting, and fanning the hive. Nobody ever described the worker bees to us as being exploited or overworked in comparison to the queen or drones, alluding to the "natural" human gendered division of labor in the private sphere and how it is believed to benefit the greater good of the individual and social family.

Many informants expressed deep appreciation for the labor of the worker bees. Sarah, a Bronx beekeeper in her late twenties, spoke passionately about her bees: "I mean, aren't they just amazing and incredible—the living breathing hive? There are many times when it is dusk and I will just watch them all come in and it is amazing how many of them all come in at once and they find their little landing pad and they head in. Their legs are all full of pollen—it is fascinating and beautiful." And it is with a deep respect of the worker bees' industrious nature as well as the acknowledgment of how bees straddle the private and public spheres of organic life. Mary Woltz, a beekeeper who has trained many newly minted five borough beekeepers, told us, "I have difficulty considering my girls pets. They are far too independent. Through their generosity they allow me into their lives/homes, and I absolutely don't take that for granted. I think one of the more intriguing aspects of beekeeping is working on the edge of domestic (which they aren't) and wild (it is an increasing challenge for them to live without us these days)."

Descriptions such as these conjure an idealized type of communal living in which everyone's job is essential as related to the greater good of the whole. While interviewing beekeepers, several who shared their admiration for bees' architectural ingenuity and "green" lifestyle, images of pod-type living in some postapocalyptic, floating self-sustaining ship came to mind. Commenting on this preoccupation with bees' talent as engineers, the architectural historian Irene Cheng has written, "Architects have long suffered from animal envy" and since the 19th-century architects have marveled over bees' "divine design."[18] It is almost as if the bees' way of life is a utopian model, an "arcology," that humans cannot achieve. And although unachievable, it sets a standard of sorts. An "arcology," a combination of the terms "architecture" and "ecology," most often used in science fiction, is a self-contained, high-density living environment that forms a fully functional residential and agricultural habitat with minimal waste. The respect shown for the efficient and egalitarian hive is an example of how the natural world seemingly "gets it right." There is an implicit narrative here—we can learn better living through bees. The teeming metropolis of New York City, with its green initiatives and urban homesteads on rooftops, attempts to repurpose dumpsters as swimming pools, bathtubs as garden planters, and found objects as art. This optimal use of space and raw materials is similar to the bees' use of the comb cell for food storage, rearing of young, and gestation of queens.

Eric Rochow, a beekeeper with a popular podcast on gardening and urban homesteading, describes the hive in the following way: "It is just a big-ass machine and it is amazing how they work. I think it is that the bees are one-cell animals that represent a single entity, the hive. Because it is all they have to be there for the thing to work. I mean, it can't just be the queen and a couple workers it has to be thirty thousand workers and a few drones and a queen. And they all have different jobs and [especially as] they age. The last week they are the foragers and then they are done." This conceptualization of the hive as machine brings to mind this arcology where all actions and movements are creating ideal living conditions and each individual function in sync for the greater good of the whole. We also must consider the capitalist work ethos/ethic and how it might trickle into this type of hive worship. For example, as seen in the American ideologies about an honest day's labor or hard work paying off, or even World War II propaganda, whereby Uncle Sam says we can do it for the cause if we all pitch in, American ideology adheres strongly to ideas about working hard. Being a good American entails an ideological belief (even if the actual practice is less evident) of a sacrifice for country and a purpose larger than the self. An honorable and patriotic existence means working hard for the greater good of the American people, our dominance in the global marketplace, and our reputation as a "can-do" nation. The idea of what it is to be a good American worker is then transposed onto the bee, and the key to having a "good society" is found within the hive. These beekeepers admire the bees' industry and self-sacrifice; their very eusocial existence appears to demonstrate a desire for social cohesion.

Bees exemplify the idea of a "superorganism," a term used by scientific researchers to describe a social unit of animals with a highly individualized division of labor wherein members cannot survive long outside of the group. According to the historian Charlotte Sleigh, the idea of the superorganism has made its way into popular culture, "inspiring the popular enthusiasm there is for myrmecology [the study of ants] as well as being reconstructed to suit a number of other fashionable topics," from Internet memes to bees.[19] Opening up the hive box and seeing this working machine of industry and seemingly sacrificial toiling represents a version of the superorganism, a community of sorts. This community made of unselfish insects is something that we yearn for in our collective imaginary and we want to replicate it as a curative for what might be lacking in urban human life. But importantly, the superorganism is a human construct and there is much debate over whether or not it exists in ant worlds or elsewhere.[20]

Almost in direct opposition from the negative and dismissive expressions toward the drone, worker bees are also able to establish affective ties with their human caretakers. B.J., a backyard urban homesteader, related to us, "I am very sad when they die. Even little ones, one was in the kiddie wading pool and I was like, 'Oh, I will save you little girl,' so I just scooped her up and took her out. And she lived." On many of the hive checks we performed with beekeepers, we observed humans gently ushering workers away from dropping frames. Hive brushes softly encouraged worker bees to safer locations with whispers of "over here, ladies, I don't want to hurt you" or "that's a girl." Clearly interested in respecting the worker's life as well as disrupting her routine as little as possible, beekeepers seemed to respect individual worker bees as highly valuable to the hive.

Beekeepers learn from the bees on many different levels; they see them as "model insects," sacrificing team players, and members of a greater collective good. By anthropomorphizing bee behaviors, people also re-learn (or reproduce) social narratives that support the status quo, stratifying humans according to gender and sexuality. This is another way in which bees are paradoxical; we embrace them as a species that holds the key to a more collective and progressive form of sociability and an idealized work ethic—a model for humans to learn from. But at the same time humans personify bee behaviors in a way that reifies gender stratification and the assumed naturalness of masculinity, femininity, and heterosexuality as shown in the beekeepers' narratives and the lessons learned from Farmer John at the Sunflower Academy. The preschoolers caught the buzz, but exactly how will they interpret it for themselves?

At pickup after preschool, Greta is thrilled to hand Lisa Jean her "bee"—a glued stack of oval-shaped paper with stripes of yellow and black paint. Other children hand their bees to their moms or nannies and gather their jackets and lunch pails. We wonder what has seeped into their consciousnesses about the species and how a whole lesson plan has been designed around a bug. Some of the students have likely left with a better understanding of the anatomy of bees and beekeeping practices: perhaps the bees will be encountered as more familiar and less scary when they come across one in their yards or at a park. But what did these kids inadvertently take away about humans as a species and themselves as burgeoning individuals through this human/insect lesson about gender socialization? Will narratives of bee heterosexuality linger in their imaginations as they wonder about the bee "mommies" caring for the "babies" after the bee "daddies" die?

It strikes us that in some ways the stories of the honeybee cleave to particularly American stories of success and failure. The element of gender and gendering bees both affirm gendered experiences of the division of labor—wherein women are still responsible for most of the private, domestic, and household tasks of reproducing, feeding, cleaning, and caring for young—as well as a protofeminist culture in which female worker bees are the foragers, protectors, and leaders of hive organization and relocation (in the case of swarms). A majority of our informants animatedly recounted stories about successful reproductions of European-immigrant hybrids ("good brood") that lead to productive ascetic workers, the female labor reserve "doing what comes naturally" and maintaining thriving colonies. These worker bees intensely focus on the survival of their species: they are good citizens, self-sacrificing nannies that are undercompensated, and making the "nation" strong. The qualities of worker bees, in particular an ability to adapt to environments that may be unwelcoming or foreign, are comparable to the characteristics of the model immigrants who first arrived in Ellis Island. The next chapter looks in greater detail on these migrations of bees and how swarming taps into human fears and desires centered around race, ethnicity, and nation.

6

Breeding Good Citizens

All-American Insects

Swarms

In spring 2011 in New York City, tens of thousands of bees collectively decided to vacate their hives in search of a more amenable place to live. Flying together in mutable bunches that can resemble a revolving insect tornado, swarms of bees ended up at a BP gas station in Brooklyn, a yellow barrier on the Lower East Side, and a mailbox in Little Italy (just in the month of May alone). When this happens in New York City, the police are first responders, areas are taped off (resembling crime scenes), and portions of city blocks become inaccessible for hours—perfect ingredients for a traditionally salacious news story. One Manhattan online news source reported, "Massive Bee Swarm Shuts Down Little Italy Corner,"[1] while another site exclaimed, "Beelieve It or Not, NYC Has Its Own Bee Rescue Team."[2] According to a NYC beekeeping codirector, Jim Fischer, "Swarming is a normal and nonaggressive behavior that typically happens during a several-week period in the spring when an older queen bee leaves the colony with a large group of worker bees to find a new home." Most would agree that "these flights are an awe-inspiring and natural means of reproduction of honey bee colonies."[3] While swarms are an awesome natural occurrence, they make for a media event, especially in a metropolitan environment where the daily presence of bees is eclipsed by roaches, lice, bedbugs, and other insects that regularly demand human attention and intervention. In figure 6.1, taken in June 2011, an NYPD beekeeper, Tony Planakis, climbs a fire-truck ladder to capture a swarm in Chinatown. This swarm, five pounds' worth, according to Planakis, shut down Mott Street for several hours while the bees were taken into custody.

When a swarm occurs, local TV news reporters, bloggers, and inquisitive crowds swell the scene with curiosity and fear. Volunteer

Figure 6.1. An FDNY ladder bee-swarm rescue. (Photo credited to DNAinfo.com/Patrick Hedlund)

urban beekeepers dressed in full protective gear, some members of self-described rescue teams, swoop in. Hooded, veiled, and gloved, looking like futuristic hazmat specialists, they swiftly coax the bees into a box and transport them away. In a few hours the spectacle is over. It's a relatively harmless and spontaneous show that plays out like an amalgamation of the Nature Channel and a sci-fi B movie. Honeybees swarm every spring, regardless of where they live—they can exist in a man-made hive box in the suburbs or in a feral untended spherical mass on a tree limb. From some beekeepers' perspectives, swarming is a positive sign but also a threat to be managed. It means that the hive is so robust and healthy that it has outgrown its original home. Swarming is a form of reproduction that some argue is a biological imperative protecting the greater good of the hive. Many beekeepers describe swarms as magical, and for those who haven't seen one, hoping to see a swarm can be likened to a backyard astronomer's search for a comet or shooting star. Both of those are fairly regular and normal natural events, but they are elusive enough to humans to render them spectacular. They are the objects of legend, anecdote, exaggeration, and art.

Swarming creates fear, as illustrated in the examples above, and whips humans into a frenzy of coordination. The anxiety about swarms is no doubt amplified by the inability to ask this alien invasion to "take us to your leader" for quick resolution or military action against whomever is in charge. Bees can relate to one another and fly through space in great numbers of 15,000 and although they do not typically sting when swarming, they are capable of it. They all collectively maneuver to the same place, staying in a moving, writhing, and buzzing collective—and these movements are not predictable by humans. It is quite astonishing—particularly when you consider how hard it is coordinate "spontaneous" flash mobs, a trendy form of public assembly where hundreds or thousands of people perform a dance routine or ride the subway without pants.[4] Those events are clearly organized, rehearsed, and orchestrated by one person who transmits instructions to other individuals.

The philosopher Eugene Thacker finds the concept of the swarm to be productive in theorizing about nontraditional forms of social organization. In his estimation, using skills of animality, or the human capacity to *think the animal*, reminds humans of their deep interconnections with other species. Thacker proposes that a swarm is an animal multiplicity where the collectivity exists as "an organization of multiple, individuated units with some relation to one another." He continues by saying that the "relation is the rule in swarms" rather than some centralized individual ruler.[5] Understanding swarms as collective organizations and not just masses or crowds, swarms are an alternative to modern sovereignty, or the rule of law typically used by a ruling government over a geographical area. Generally, in sovereign rule, there is a centralized control that decides for the rest of the group. However, in swarms, there is no individual decision maker; instead there is a collectivity "defined by relationality." Cognizant of this as a paradox—that swarms have purpose but no overarching director—Thacker believes that swarms are confounding in that they demonstrate social action without a ruler.

From some of the earliest written records, humans' observations and interpretations of bees' activities, colony construction, and hive behavior have served as fertile ground for understanding and enhancing human domination and social organization. Featured in Virgil's, and subsequently Aristotle's and Plato's, writings and poems, bees are used as a cross-cultural metaphor, particularly to comment on Roman imperial regime building.[6] Bees are situated within violence yet imbued with a moral purpose. Bees have been useful to humans as weapons, radar, and even robots. The anthropologist Jake Kosek has even gone so far as to argue that honeybees in the 21st century are "sensory prostheses" for human military interests.[7] In that way honeybees are recast through technological innovation as instrumental agents ready for human use.

Swarms in particular appear to be easily ready for combat, whether in theory or practice. But outside of military rhetoric, a swarm can be viewed as neutral and harmless. For example, according to expert beekeepers whom we interviewed, bees swarm when a group splits from the hive to create new hives, enabling their species to grow so the colony can reproduce itself. Reasons for swarming include overcrowding, illness, dying or aging queens, and temperate winters when food supplies run low. A swarm of bees from the perspective of a beekeeper is a nonviolent "natural" occurrence. Similarly, a radical ecocultural feminist could interpret bees as the perfect holistic and peaceable kingdom of cooperation and matriarchy. The meaning of swarming from this theoretical standpoint is based on reproduction and sustainability—the swarm just takes half the hive and leaves to make or find another queen. In this chapter we draw from the many different interpretations of the swarm, as part of the buzz—as an unpredictable mass that creates fear and hysteria, a type of collective social organization, and an instinctive occurrence or natural state of being—as we continue to follow the buzz around bees, moving from gender and sexuality to race and ethnicity. By humans, bees are characterized as "social" insects but they are also uniquely *political* insects. Within U.S. borders, certain bees are more welcomed than others.

"Your huddled masses yearning to breathe free": Immigrants

Almost all of the research we conducted took place in the New York metro area; however, in the middle of our project in the summer of 2011, Lisa Jean taught a food studies class in Southern Italy for students from the State University of New York. While there, she interviewed beekeepers. We were curious to compare beekeeping practices in such diverse locales, and we looked forward to learning about the Italian honeybees and their keepers. Going off course geographically and culturally (while staying on topic) provided a reorientation to the project. American honeybees, like the majority of American citizens themselves, have ancestors that originally came from someplace else. As we have come to understand, the bees' place of origin is also highly significant.

Bees in Southern Italy

Eighty-year-old Aniello Piullo lives in Pisciotta, an Italian town south of Naples on the Cilentan Coast. His home, which he built fifty years before, is an incredible example of self-sustainability—though without the self-consciousness of the urban homesteader. He and his wife grow most of their food and raise chickens and goats. They make their own wine grown from their vines, press their own olive oil, and harvest their own honey. Ripe eggplants, tomatoes, and squash dangle along the walk up to his doorway. His stone deck is covered in fleshy apricots and plums that stick to our shoes. Despite the ominous sign on the door "Non svegliare la nonna e il nonno" (Don't wake up Grandma and Grandpa), the Piullos wait for us with espressos on the side terrace. Mrs. Piullo wants me to taste her chestnut honey before we go on a tour.

Once fueled with espresso and honey, we are led up a treacherous cliff through winding switchbacks and dried bramble to a terraced side of a mountain. As we approach the clearing, there is an audible hum and a familiar slight sweet smell wafting through the olive trees. Opening the branches of a tree with his cane, Mr. Piullo guides me to the fifty to sixty wooden hives placed on rebar, cinder blocks, and terra cotta (see figure 6.2).

Snapping pictures, Mr. Piullo tells my translator that he has never had anyone photograph his hives, and he chuckles, amused at my scrambling up and down the terraced levels to get good shots of the bees on their landing strips. Focusing in on the furry creatures, I am uncharacteristically awestruck: "These are the real, authentic Italian honeybees—at the source." Having met "Italian" honeybees in Brooklyn, I think about how these bees might be in some way related to the bees I saw in New York City. And in my wonder at the glory of the

Figure 6.2. The Piullos' Italian honeybees perched on a cliff overlooking the Mediterranean. (Photo credit: Lisa Jean Moore)

authenticity of "discovering" them in their natural habitat, I marvel at how they seem somehow softer, fuzzier, and more vibrantly colored than their American imported sisters. Clearly I am romanticizing the bees but I am swept up. These bees are flourishing, native, and free, almost happy in their self-determined work in this beautiful Eden (see figure 6.3).

Reflexively, it is clear I am caught up in the grandeur of the countryside, the Italianness of the bees, the unadorned beauty of the Piullos' lifestyle. Thinking back to Brooklyn I can't help but compare my bucolic surroundings with the urban homesteading and beekeeping movement back in NYC. In this light, rooftop farming seems inauthentic and manufactured—part of a fad, or lifestyle, affectation, politics, purpose, identity, or hobby versus living the life of the Piullos. This lived everyday experience was slightly dirty and messy, albeit beautiful, but it was clearly not aesthetically maintained. Obviously this romanticized "real deal" where there are Italian honeybees living close to the source and their "origin" amplifies their status as the ancestors of the American immigrants. And there is the seduction of nostalgia that enhances the genuineness of the Piullos' beekeeping. Their genuineness suggests that there is an oddly fabricated/phony self-consciousness about the people

Figure 6.3. The landing strip for the Italian honeybees. (Photo credit: Lisa Jean Moore)

Figure 6.4. The Piullos with a large jar of their honey. (Photo credit: Lisa Jean Moore)

in NYC who do urban beekeeping. But I was doing some cultural work myself: seeing the Piullos and their bees as somehow more real than the American counterparts (see figure 6.4).

Coming to America

All honeybees in the United States are from imported relatives. The honeybee is not native to the Americas but rather is a diasporic species. Honeybees were brought to North American environments in the 17th century and integrated into the American colonists' beekeeping practices.[8] There were upward of four thousand other species of bees native to North America, prior to their arrival. As of 2009, the U.S. Department of Agriculture (USDA) estimates that there are between 139,000 and 212,000 beekeepers in the country, mostly composed of hobbyist beekeepers. There are roughly 2.68 million colonies. According to the National Honey Board, on average each individual American consumes 1.3 pounds of honey per year.[9] In 2008, the United States produced about 160 million pounds of honey,[10] though much of the honey we consume is imported from other countries.

Honeybees are not immigrants per se but they do have a migration story, which is comparable to the story of certain types of Continental immigrants. Because insects are so tiny and often elusive, we rarely think of bees within larger social networks or systems, certainly not as nonhuman members of a nation-state. Bees were originally "let in" through U.S. customs, as it were, without any restrictions. Over time, and for the greater good of the populace, bee migration became an issue of national security. National borders are supposed to regulate human traffic and in doing so they construct different nations, states, and races. Humans, as well as animals and insects, are also constructed as foreigners in order to shore up the notion of nation. As we argue, similar to the rise of the modern state that polices its physical borders, honeybees are completely implicated in and part of the story of race, nation, and difference.

Relatively unregulated for a couple of centuries, it was not until 1922 that the United States established regulations in order to protect against pests, viruses, and diseases. The key legislation is the Honeybee Restriction Act of 1922, which gave the USDA the ability to regulate honeybees. The USDA's Animal and Plant Health Inspection Service (APHIS) has regulated the importation of honeybees since the 1922 act. The Plant Protection Act of 2000 regulates the broad group of pollinators including honeybees, native bees, butterflies, and moths. Current law in the United States stipulates that honeybees can be imported from only New Zealand and Canada. Interestingly, much of the legislation about honeybees has been about limiting the entrance of these bees into the United States, instead of investigating how to maintain the health of the bees that are already here. Another notable bee-related U.S. legislative act was the Beekeeper Indemnity Program of 1970 administered through the USDA until 1977. This program compensated beekeepers for the loss of their colonies when they had proven exposure to federally approved pesticides. Partially in response to Colony Collapse Disorder (CCD), the Farm Bill of 2008 included specific funding mechanisms to encourage farmers to enhance bee habitats on private farms and ranches.

We present these regulations in part to illustrate the ways in which what we would call "biopolitics," or how governments control organic life, are at play in the existence of the honeybee—through these legislative acts, the life of the honeybee is both fostered and prevented.[11] The USDA regulations are designed to allow imports of honeybees from only countries where the bees have specific diseases that are already present in the United States and to allow the import of bees from only countries that have similar regulatory structures to the United States. The regulation involves an active screening of honeybee health in other nations—as indicated by one official's explanation that "we had let imports from Australia in until last Fall [2010] but then the Asian Honeybee problems became too significant for us to let in those bees." Asian honeybees, *Apis cerana*, hail from southeastern Asia, including China, India, Korea, Japan, and Malaysia.[12] These bees are the preferred host for *Varroa*. The coevolution of *Varroa* and Asian honeybees means that these bees have grooming behaviors that eliminate the devastating effects of infestation.

However, European honeybees do not have this same skill and fall prey to these mites. Policies in biosecurity, particularly in Australia and New Zealand, have attempted to manage the Asian honeybees. In the United States, Australian bees are now illegal to import because of unregulated border crossing from Asian countries to Australia, leading to new hybrids. The USDA has a specific agenda to keep out the Asian honeybee—but not the Asian honey. In other words, there is no free flow of labor across borders, but there is a free flow of goods—think NAFTA. With bees, there is no free flow of labor across borders, but there is a free flow of product. Ultimately, just like human workers, nonhuman workers are bound within the policies and borders of the nation-state. Consumer goods, because they have economic exchange value, circulate more freely in the global marketplace between countries and across continents.

Even though the health of the honeybee is being "managed" by legislation across borders, the health of humans who consume honey in the United States is another matter. More than half of the honey consumed in the United States is imported from other countries, including Argentina, Brazil, and Mexico, but the majority comes from China. While it is difficult to determine how much honey is of Chinese origin because the distribution practices are aimed at evading inspection,[13] Chinese honey is cheaper than domestic honey. Honey laundering, or faking countries of origin in order to get Chinese honey into the United States, has led to an indeterminate amount of tainted honey—honey that contains sugars, heavy metals, and illegal antibiotics—entering the market. This practice, in global trade of all products, is called transshipping. Senator Kent Conrad of North Dakota, the number one honey-producing state in the Union, has stated, "Unfortunately, China has a long track record of importing adulterated honey and engaging in other fraudulent conduct in the honey trade. These actions not only hurt honey producers in North Dakota and across the country but also present needless health risks to our consumers."[14] He goes on to state the fears about the extent of Chinese practices:

> As shown on the FDA's on-line listing of import refusals, the growing number of import refusals for impure, adulterated or otherwise unfit products from China far exceeds refusals for other countries. We fear that these reported incidents may only be a portion of a much larger problem. We are particularly concerned about common practices that may enable those who adulterate or mislabel imported honey to readily escape detection. For example, the continually changing list of

> enterprises selling honey from China, and the extensive history of fraud and illegal transshipment in honey imports from China may make it especially difficult to determine the actual producers of impure imported honey.[15]

Furthermore, recent testing of honey in American supermarkets revealed that approximately 75 percent was ultrafiltered, a practice in which water is added to honey and it is filtered to such an extent that it is no longer considered honey and it is devoid of pollen.[16] Pollen is the only way in which human testers, called melissopalynologists, can determine the origin of honey.

Asian honeybees and their human counterparts are suspect for their diseases and their companion species' (i.e., humans') disreputable practices. However, it appears easier for the U.S. government to establish mechanisms to disallow Asian bees entry into the country, while humanly altered honey manages to sneak into our borders without too much difficulty. Honey is a global commodity, and its sale and distribution equates into hundreds of millions of dollars of revenue. Given the popularity of honey, the dietary demand (and profits) will likely veil potentially pernicious honey supplies and the unscrupulous nature of the trade. Bees as invader insects make for a clear target on governmental radar, but the product of their labor is a form of currency and, as such, is open to obfuscation within a market economy. The black-market Chinese honey trade[17] gives new meaning to the well-worn adage that honey is "liquid gold."

Breeding Better Bees

This global trade in bee products is often not the primary concern of the urban beekeeper, although the organic and locavore food culture certainly undergirds many urban beekeepers' practices and philosophies. Beekeepers are most concerned that their honeybees are protected from adverse weather conditions, diseases, and parasites, such as *Varroa* and tracheal mites. Weather, disease, and parasites can devastate a hive and a great deal of the work of beekeeping focuses on providing environmental protections against these risks. One way to manage risk is through selecting a subspecies of honeybee to raise, usually beginning with the queen and the package. Each of these subspecies has its own migration story and particular characteristics that are said to differentiate it from other types of bees. Much like humans differentiate between races and ethnicities, according to geographic origin, bees are also stratified and regulated. Bees are subject to eugenics, depicted as valuable contributors to the national ecosystem or in some cases considered to be dangerous "illegal" aliens.

Among apiarists, *Apis mellifera mellifera* is the Linnaean nomenclature used to identify European honeybees. The more informal taxonomic rank of race is below that of the subspecies of bees, beekeepers categorize specific differences as "racial"—not ethnic differences, a term we use in describing human beings. The three most common types of honeybees in the New York City area have made transatlantic and cross-continental journeys, hailing from Italy, Eastern Europe, and Russia. *Italian honeybees* (or A.M. *ligustica)* have been around since the late 1800s and are the most popular honeybee in the United States, while *Carniolan honeybees* (or A.M. *carnica Pollman*), natives of Hungary, Romania, Bulgaria, and Slovenia, are the second most popular subspecies in the United States. The most recent "immigrant," *Russian honeybees* (or A.M. *caucasica*), are fairly new to the United States and are thought to have greater resistance to mites. Our informants describe them as particularly hearty even when living under less than ideal conditions. These types of honeybees also interbreed and create a resulting hybrid bee. According to a USDA official we spoke with, "The European subspecies mix well together." This work of establishing pedigree is another frame (political, cultural, racial) through which we differentiate certain species and exercise control over nonhuman others, in this case, for our own interests.

In conducting our research, we continuously resisted the urge to believe there was a utopian bee world once—a time when honeybees were "pure"—as suggested in the fieldnotes from Italy from earlier in this chapter. There is a tacit idea that Russian or Italian honeybees just spring forth fully formed from nature and that these bees did not have some type of manipulation prior to the rise of the modern nation-state. We have worked to avoid the nostalgic notion of the "original" honeybee as it lingers in our imaginations.[18] Animal and insect husbandry is not new; it is, however, possibly more obvious and widespread with degrees of specification that are unparalleled in history. Like much of natural life, including the lives and bodies of humans, organic matter is increasingly technologically manipulated. Incredibly, we genetically manipulate gametes (i.e., reproductive cells) from several species, including plant life where we have lost the notion of indigenous or native seeds and human life where we are able to prevent serious genetic disease.[19]

When humans stereotype and create simplified generalizations about an entire group of people or individuals based on their perceived membership to a group, they are reproducing and reinforcing layers of stratification. Stratification, a sociological term, refers to the hierarchical organization of groups into levels of relative power and social worth. The majority of the urban beekeepers we interviewed and observed were urbane middle-class, white people. This demographic group, who has been educated since the massive social protests of the 1960s and 1970s, would seem to be quite well versed in resisting prejudicial racial/ethnic or gender characterizations of honeybees. In some cases, our informants have chosen beekeeping as some form of countercultural resistance to hegemonic forces, particularly with regards to food consumption and commercial capitalism. Thus, it was sometimes startling for us to encounter cavalier racial remarks and jocular commentary with regards to bee interbreeding: "Italians breed a lot and then eat all their stores—not very smart" or "Killer bees are known to be ferocious." Racial politics and breeding practices somehow escape the politically correct worlds of some educated and presumably multicultural beekeepers.

Stereotyping is practiced by humans over other species through, for example, representations that anthropomorphize and negatively generalize: pigs become lazy and filthy, lions are noble and royal, and bees are busy and industrious. Beyond this type of projected cultural inscription, beekeepers are quick to point out the race of their bees and to attribute certain qualities or behaviors in association with this race. Italian honeybees are thought to be more docile, "the most calm bee" and passive, great reproducers of brood, and easy to work with. As Gita states, "My ones last year were very docile. They were Italian and I felt like we had a really good relationship, I did get stung a couple of times but they never, like, attacked me." The ability for her to have a positive interspecies relationship depends on bees being docile, predictable, and tamed. And as Gita describes, her bees' disposition is in some way connected to their Italian heritage. Cast in this frame, Italian honeybees are the model minority—the original hardworking immigrant that is wanted and desired, assimilates, and does its own work.

However, Italian bees are known to "greedily" eat their stores of honey or even become honey stealers from different hives, perhaps because, as one beekeeper put it, "they need a lot of food." Eric Rochow, a beekeeper for more than four years, notes, "Italians are a gentle and weak bee. Russians are better, especially where I live because it gets really cold up there." Russian honeybees are characterized as vigorous and good at surviving harsh winter conditions; however, their heartiness can give way to aggression. This aggressive behavior or the fear of it can lead beekeepers to discriminate against this type of bee. Eric explains how the genealogy works: "Because these are Italian bees, Italian queen, Italian bees." He goes on to describe some racial differences: "A Russian, except that they are certainly hearty, but they have a greater tendency to swarm in the summer. And they can be aggressive. They are really hearty." In his explanation of racial differences, Eric also revealed there are trends and preferences: "But there is a hybrid I might get. I think that people do like designer bees and 'oh I have the Russians'—I mean, good for you, that's nice." He smiled and shrugged as if to say "these crazy people" while at the same time admitting that he participates in the racial selection process. Among our informants, racializing bees was largely taken for granted.

Carniolan bees are often depicted as "great honey producers," and many beekeepers take pride in the amount of honey their hives yield. It is tangible evidence of hive health and good beekeeping practices—a symbolic payment and trophy for a job well done. Furthermore, hybridization using Carniolan stock is explained as beneficial. Beekeepers spoke to us about selective breeding toward maximizing the desirable qualities for using bees in colder climates. Meg Paska shared her success with hybridized bees: "My queen on the roof that overwintered is an Italian Carny hybrid. That's good. I have also tried Russian bees. Everyone warned us about their temperament but it was no different than the Italians." Some beekeepers are willing to experiment with new breeds, albeit keeping in mind the characteristics that certain breeds are commonly known to have within beekeeping cultures.

Racial differences can also be expressed to demonstrate that the bees behave in ways that will jeopardize their survival—implying that the beekeeper needs to be aware of some of the racial characteristics. In this passage, a beekeeper describes the differences between Russian bees and Italian bees. "The difference between a Russian hive and an Italian hive—Russian hive is a small basketball-size cluster, no brood, bees ready to winter and maybe fifty pounds of honey to metabolize over the winter and the meantime during the same time like in November you will look in the Italian hive with the Italian queen and it will have four or five combs of brood, filling the boxes, a huge population, and they will be totally starving." He went on, in a way almost irritated at the bees for not thinking ahead: "They will turn all their food into brood, expecting another honey flow to come. They maintain a larger cluster for the sporadic honey flows that are unpredictable. Not the best planning."

As described, he laments how Italian bees are not able to manage their reproduction with the availability of food. How much do these descriptions of Italian honeybees mesh with stereotypical understanding of Italian people? As illustrated through the comments of the beekeepers, there is as much bleeding between ethnic stereotypes about human ethnic groups as there are about ethnic bees. Italian bees are happy and contented and gorge themselves. But they potentially eat themselves out of house and home and push their reproductive practices well beyond their ability to support themselves. The libidinal lifestyle requires more maintenance and intervention from the knowing and surveilling beekeeper. In juxtaposition, Russian bees are hearty in a seemingly Soviet-style approach to survival through disciplined and austere conditions and their commitment to the hive. Their strength and toughness require less intervention on behalf of the beekeepers and they are admired for their strength while there is a slight sacrifice at the closeness and intimacy that can be shared between human and insect.

Because we are not expert beekeepers or entomologists, we are not equipped to determine whether the racial traits ascribed to bees are verifiable. We are not interested here in figuring out how whether or not these stereotypes are true, and we doubt it is even possible to determine if they are objectively true. Instead, we are interested in how humans narrate bee ethnic differences. It is possible that beekeepers that already know the ethnicity of their bees read into their bees the already extant human-bee ethnic stereotypes. As sociologists, we are interested in how beekeepers seem to uncritically and unconsciously use human stereotypes to differentiate certain bees. The racial taxonomy of bees often mirrors assumptions about human racial differences.

Russian bees are bred for mite and disease resistance and peak performance in a variety of climates. Imported in 1997, through a special program of the USDA, one hundred Russian queen bees were permitted to come into the country because of their noted parasitic resistance. These bees were then quarantined off the coast of Louisiana and determined to be "safe" only after they were moved to a laboratory where they were subjected to a battery of breeding cycles, mite resistance tests, and trait analysis. The eugenic program of enhanced Russian honeybees was monitored by apiarists to breed bees specifically for mite resistance, winter readiness, and honey production. These bees were then carefully sold through existing apiary networks. As explained by the USDA, "Under the agreement, third-generation apiarist Steven S. Bernard is authorized to raise and sell pure-Russian breeder queen bees on a first-come, first-served basis. The breeder queens cost $500 each. From each of these, beekeepers breed thousands of production queens, which are placed in hives for pollination and honeymaking. Strict mating control of production queens is not done, so they sell for about $10 to $15 each."[20] Similar to the sensationalized reports of Soviet athletes being enhanced through technoscience, Russian bees were indeed tampered with to provide a "new and improved" species for human deployment and super performance. However, this genealogy is obscured by fifteen years of integrating Russian bees into the general honeybee population.

Although Italian honeybees have a high status among beekeepers, some of our informants preferred Russian bees. Sam Comfort, whose previous experience included working in commercial bee yards, is now committed to organic, treatment-free beekeeping and has a preference for Russian bees or feral colonies he collects through retrieving swarms. He attributes the heartiness of the Russian bee to the lack of centuries of manipulation of these bees, primarily because they have not been in the United States for very long. In Comfort's estimation, Russian honeybees do not have the same level of assimilation of Italians or Carniolans:

> Russian bees are resistant to everything—gut diseases, secondary diseases, they do really well with little management, winter well, take care of themselves with no management, they swarm like crazy and are basically like wild bees. And meanwhile all the other bees in this country can't compare to them . . . they are like cream puffs. They are just not competitive. And I just think it is because the Russian bees haven't been messed with like other bees. I just see similarities between the Russian bees and the wild bees or the so-called African bees. They all have this spunk. When I investigate, I see how the wild bees do the best, the other bees just couldn't break cluster in winter.

Because most beekeepers must manage the conditions to prevent swarming, a reproductive practice that creates a new home for a portion of an existing colony, bees that are more likely to swarm are troublemakers. While many report the spectacle of a swarm with animated descriptions—"It was fantastic," "It was a religious experience," "They are so impressive and calm the way they swarm" —most humans very assiduously work to avoid them. Hence, this explains the preference for Italian honeybees, which are considered easier to manipulate. A small segment of beekeepers, particularly those from the more naturalist or self-described backwards community, considers the swarm to be "good, wild nature" or nature as it should be. Comfort and prominent L.A. beekeeper Kirk Anderson, who advocated radical *dis*-intervention, believe that humans should "let bees be bees," referring to the idea that bees should swarm on their own and not be "forced into living together" by humans. This hands-off philosophy has attracted a strand of beekeepers within the broader green movement who adhere to notions of human intervention as something to be managed with respect to bees, versus the management of the bees themselves.

During our fieldwork at Comfort's Beekeeping Bootcamp, we visited a bee yard that featured hives from captured swarms. Clearly proud of these hives and their presumed "feral" nature, he placed them slightly apart from the other boxes and pointed them out to us in more revered tones: "Those bees I got from a swarm out in the woods. They are a feral colony. They are survivors." For these beekeepers, the feral bee colony is akin to finding a wild horse or wolf puppy—ironically something to be captured and yet able to retain the status of "wildness"—outside of the puppy mill, horse breeder, or apiary. So instead of the discourse of protecting nature including the bees themselves, there is a discourse of liberating nature. This spirit is embodied in a more radical and masculinized beekeeper stance that seems to capture the renegade spirit of the deep ecology movement. As the philosopher Michael Zimmerman explains, the deep ecology movement, which emerged in the 1970s, requires that humans move away from *anthropocentrism*, or seeing their species as the center of everything.[21] Rather humans should see themselves as connected to the whole ecosphere, where injury to anything within the environment is injurious to all living things. Akin to ecofeminism, a version of cultural feminism that embraces women's spirituality and relationality to the natural world, deep ecology appeals more to renegade beekeepers who emphasize the wildness of their bees.

Killer Bees: Fear of the Dark Swarm

In the late 1970s, media reports warned Americans of threatening Africanized honeybees that were more aggressive than the European honeybee. Africanized bees (or *Apis mellifera scutellata Lepeletier*), visibly similar to their European cousins, are smaller honeybees that originate from tropical Africa. Also, in comparison to European bees, Africanized queens lay more eggs, which develop more quickly into adult bees.[22] As reported by the media, a veritable buzzing, swarming "other" was coming to get us, originating in Africa and migrating from Brazil through Mexico. The Africanized bee was in many ways framed as an illegal alien: unfamiliar and unwanted, exploding in population and heading straight for our borders. In this case life imitated art, as the popular 1974 novel *The Swarm* and subsequent 1978 film of the same name used Africanized "killer" bees as a Hitchcock-inspired protagonist. This exotic dark villain was depicted as devastating farms, attacking people, and eventually ominously darkening the New York City skyline.

The USDA tracks the migration of the Africanized honeybees, which first arrived in the United States in Texas in 1990. See, for example, figure 6.5, taken by our colleague hiking in Arizona. These aggressive bees require that a warning label be provided for the harmless and defenseless human climber who is out to enjoy the pristine desert landscape. Notice that the bee's body is drawn to human scale (with a sharp

Figure 6.5. Signage at the beginning of a hiking trail in Arizona.
(Photo credit: Matthew Immergut)

tooth-like stinger), making it appear especially dangerous. In addition, the man has his back turned to the bee as if to suggest that Africanized bees will creep up and strike without warning or provocation. This sign sends a clear message—these insects are among us and are unpredictable and to be feared.

The entomologist Mark Winston points out that the term "Africanized" bees is a wholly American phenomenon of naming. As he traces the genealogy of the Africanized bee, Winston performs a content analysis of major media headlines from the 1970s, including references to the bees as "monster creations of science fiction," "possess[ing] a rage," and "savages."[23] Africanized honeybees are a subspecies of bees whose swarms are capable of overwhelming a hive of European honeybees, killing the queen, and superseding her with an African queen. In the 1950s and 1960s, experiments with African bees led to interbreeding with European bees, toward the Africanized hybrid. Africanized bees were prized for their high honey production and yet feared for their characterization as aggressive, nervous, and persistent. Attempts to contain the African bee were unsuccessful and they "escaped" to breed with the existing Europeans in the South American countryside.[24] These bees entered the United States from Brazil through Central America. Because these bees are not able to survive cooler climates, the USDA can track how far north these bees have traveled. Currently, Africanized bees are found as far north as Georgia. Contrary to popular perception, Africanized bees do not have more harmful venom; rather, they are considered more dangerous because they are easier to agitate, sometimes create hives in the ground, attack in greater numbers than European hives, and will attack over greater distances, sometimes chasing a threat for up to a mile. Africanized bee colonies are more difficult to identify as they do not act like European bees—in other words, they establish their hives in unexpected locations underground or in sidewalls. Because they are "out of place" from a human standpoint then they are more likely to be disturbed and agitated.

Although they are not found in the Northeast, the Africanized honeybees, sometimes called "killer bees," are present in the mind of urban beekeepers as a potential threat to be managed and feared. The Africanized bee is culturally marked by North Americans and South Americans differently—instead of killer bees, they are called "abejas bravas" or "fierce bees" by South American beekeepers.[25] Even though "African bees" are mentioned in the *New York Times* as early as 1883, according to the librarian Darcy Gervasio, it is difficult to trace the exact etymology of the term. The phrase "killer bees" starts popping up in American newspapers around August 1965—and there is an increasing specter of harm that seems to generate more occurrences of the term throughout the late 1960s. Recalling our own childhoods, we remember this media hysteria, a monolithic framing story of menacing swarms, a moment in American popular culture when some learned to fear the exotic, notably hostile insects.

There are reasons why certain narratives of harm circulate and stay with us; good stories are worth retelling because they have emotional impact. National news stories about family pets (usually dogs) being stung to death by bees are frightening examples of the potential power these insects possess over man's best friend. Africanized bees, those illegally and aggressively breaching our borders, carry with them not only particular behavioral traits but also larger ideas about ethnicity and inter/

intra species' purity and safety. Even though this hybrid breed of bee has never made it to the Northeast and attacked European honeybee hives or dogs or humans, some humans are concerned about their migration.

Many of our beekeeper informants discussed the African bee "problem" and the degree to which the danger is real. Outside of beekeeping cultures, media coverage of the potential threat of Africanized bees has often reproduced a particular narrative—a pernicious species of insect is flying past national borders, state lines, and geographical areas. And there appears to be no way to stop them; they are flying foreign objects undaunted by distance or landscape. The human emotional component of this phenomenon is heightened by the fact that these insects are classified by humans as a particularly othered species—bees that are bold, savage, and unrelenting. In other words, there are certain bees that present us with a problem—a species is labeled as out of human control and potentially harmful to us. As such, we have no ethical or moral connection to these bees, so killing them seems rational. We also haven't found a way to contain the bees for decades, making their presence a long-standing "crisis." Out of this narrative arises human emotion—in a word, fear.

But some of our fear is not based on experience or empirical verification—it is perhaps more emotionally based. For example, the theorist Brian Massumi explains how Americans became afraid of "ill-localized enemies" due to political discourses that advanced the idea of the "war on drugs" and the "war on crime."[26] Later, after the September 11 attacks, the nation would wage another and arguably, more consequential "war on terror." Based on a horrific event and the feeling of terror it inspired, more than a fact became solidified, not through any evidence, but a public mood. Massumi calls this "affective facts," or the experience of making something real by feeling it even though it has not yet happened. Bee stings can be affective facts—many fear them who have yet to experience them. Massumi asserts that affective facts play a crucial role in our lives: "Threat triggers fear. The fear is of disruption. The fear *is* a disruption. The mechanism is a capacity that affect itself has to self-effect. . . . As soon as there is any sign of threat, its most feared effects have already begun to materialize."[27] Likewise, no Africanized bees have ever come to the Northeast, but people are highly concerned about their migration and quite literally terrorized.

In our training classes, we were first introduced to the notion of Africanized bees, beyond our teenage B movie exposure, by Jim Fischer's response to an audience question: "Africanized bees had more predators and are more defensive to scare the predator. The European honeybee is not as defensive or aggressive. They are more docile and shouldn't be getting defensive. So they are preferable for people to learn with—plus the Africanized bee can't survive the winter." Sara, a community educator about urban ecology, explained to us how even in the most metropolitan settings, laypeople are aware of Africanized bees: "I have heard that because of the climate they can't move farther north than Texas and Florida. But people are really scared, especially urban people and especially teens that I work with are terrified of any bee, especially killer bees. They have no concept of them as pollinators or the utility of that in our food."

As Eric related, he keeps in touch with other beekeepers and collects their stories about Africanized bees: "I am glad I don't live in the South. I know a guy who catches swarms in Florida. I follow him on Twitter. Lots of times he needs to euthanize the bees because they are Africanized. He does bee removal as they move into people's homes and he has to cut the wall out and he determines from aggressiveness if they are Africanized or not and then they are vacuumed into a box and euthanized by dumping soapy water on them. I have never seen them. . . . It scares me that they might evolve and move up to the North." The use of the word "euthanize" is interesting here as it evokes the notion of ending the bees' suffering through the compassionate administration of relief from earthly misery. These bees are being killed for being Africanized bees rather than being recovered as a swarm for domestication and honey production. What type of cultural work does the term "euthanize"—a euphemism for intentional and premeditated killing—perform for the beekeeper? This suggests that certain bees are deemed rescueworthy while other bees are let to die in a form of insect necropolitics.[28]

Hybridization, the mixing of two different racial groups, is being pursued with most honeybees; however, it is not universally desired to include Africanized bees in the genetic mix. As an official from the USDA indicated, "With the Africanized bees, they are aggressive and have aggressive mating patterns—and don't hybridize well—the way they mate and drive out other queens does not lead to good hybridization. So what you have is the Africanized genome supplants any other genome—like European. Africanized bees have small colonies, and swarm more and aren't great producers—so they don't mix well at all with the Europeans." Using the terms "genomic" and "hybridization," this official conveys how the takeover by the Africanized bees is scientifically relevant and nationally scrutinized. Clearly, Africanized bees are believed to have traits that are inferior to European species. Successful hybridization is achieved through the use of particular human metrics; bees must perform well, according to scientists. Significant here is how eugenic discourses stem from human interests in producing docile yet highly productive honeybees, not from the honeybees' "biological" drive. Human intervention, based on genetic evaluations, is the dominant framework in which hybridization is considered. The idea that honeybees themselves would be naturally driven toward mating with other species is absent from these types of rationalized discussions. What we have in mind for the bees is presumed to be in their best genetic interest.

As the anthropologist Anna Tsing argues, "The Department of Agriculture's reaction to the African bees combines fears of Mexican immigrants creeping over inadequately patrolled borders and fears of Black-White racial miscegenation."[29] In 1864 politicians coined the term "miscegenation" to refer to the mixture of two or more races.[30] As the historian Peggy Pascoe has shown, as late as 1999, miscegenation laws included within state constitutions made marriage illegal between a white person with someone with one-eighth or more "negro blood."[31] Similar to the varying historical laws of human racial mixing, humans' quest for racial purity among their own species is mirrored by some in the quest for racial consistency among bees—at least the European ones. And yet, unlike humans' failed attempts to quantify racial purity among members of their own species, doing so with the bees is unclear. These eugenic practices imply that humans must intervene to prevent "Africanized" genes from entering the U.S. bee gene pool, which presumes that this gene pool is in some way pure to begin with. However, the bee gene pool is established as a clear mixture of bees from several European countries and ethnic groups. Furthermore there is no hard and fast scientific rule that establishes how many "African" genes are required to call a hive "Africanized." As Winston states, such a "subjective taxonomy" cannot be used to tell "how 'European' or how 'African' a colony is."[32] However, there is the fact that Africanized bees, once introduced to the terrain of the European honeybees, often outnumber and overwhelm the queen. Winston goes on to comment that "colonies can change from European to Africanized through either mating or direct colony takeover, but mating is by far the more significant factor. The Africanized bees are not 'sexier' than the Europeans; rather the European colonies become islands in a vast sea of Africanized feral colonies and are overwhelmed by the sheer number of Africanized drones at mating areas."[33]

The characterization of colonial takeover brings to mind the African American feminist and activist Angela Davis's analysis of the myth of the black male rapist. Her work deconstructs the cultural fabrication of the threat of black men and then uses this threat to justify the control of black men.[34] Africanized drones, like "dangerous black men," are described as stealthily preying on European queens, which then justifies the extermination of these lethal predators. However, we know queens mate with several different drones and drones do not survive the mating.

Basing the racist practice of extermination on "genomic" science and cultural descriptions, there is no discussion among our informants as to how bee migration and movements are indeed imperatives for bees. Bees survive through reproducing colonies and swarming to new locations. How do human-generated imperatives and interventions to bee biology and reproduction, on behalf of human safety and comfort, potentially affect the bees themselves, and how have they done so historically? Because bees are vital insects integrated into the food chain, it is assumed that we can and should manage their reproduction and migration. It is believed that bees are as mutable as crops, as in the genetic engineering of corn. Changing the plant genes to make better plants that can become monocrops immune to pests, resilient to weather variations, and ready for transport across the globe is undertaken through notions like "health," "hunger," and "sustainability." The manipulation of the bee through antibiotic treatment, selective breeding, industrial labor, and habitat design has changed the species for the convenience of human consumers.

Returning to Sam Comfort, the once commercial beekeeper turned anarchist apiarist, he has much to say about the Africanized bee story. During our Beekeeping Bootcamp where we were taught to practice noninvasive or low-treatment beekeeping with "top bar hives"—hives that were used prior to the now ubiquitous Langstroth hive that became popular during the Industrial Revolution. In addition to learning about alternative forms of hives, we were able to discuss with Comfort the racial taxonomies of bees. Lisa Jean asked him, "Is there a preference for the European honeybee? Is that real?" He was very eager to speak about this topic and shared:

> Yes, but it is an interesting story. These African bees were brought by researchers in the 1950s to increase their gene pool for a tropical working bee and they escaped and started working up Central America and everyone started getting freaked out that the African Killer bees were coming. But I know beekeepers who were in Brazil in the '70s and smuggled back queens directly from Brazil, bees that were very aggressive and meant to do that same thing, increase their gene pool. There were beekeepers who were queen producers who were large established beekeepers in Florida. I talked to the guys who were just observers of this stuff—they grabbed the hottest bees—they have been here a long time. These bees were brought here specifically to breed—Africanized bees, just the queen to build better bees because everyone thinks we need better bees and not better farming.

It is difficult to find these oral histories in standard beekeeping books. But Comfort seems emphatic in his urgency to set the record straight about the Africanized queens. Attractive because of their ability to quickly reproduce colonies, Africanized bees were specifically targeted by beekeepers for breeding experiments. More bees mean more honey and more pollinators. And, as Comfort notes, instead of changing human practices of farming in ways that encourage bees' livelihood such as planting native pollination gardens, making sure there is water supplies, and encouraging the bees to overwinter, humans have been more committed to building better bees by changing their breeding. So while it is true that some bees escaped from regulated apiary laboratories, other beekeepers specifically and furtively worked with Africanized bees for decades. This alternative narrative of humans choosing to work with Africanized bees is one that has not been recorded.

As for the reputation of more dangerous bees, Comfort dispels this belief by stating, "Every bee has its own personality. There are plenty of northern hives that are just as aggressive as African hives. And actually African hives are just defending their—they are defensive; they are not aggressive." Taking the perspective of the bee, Comfort goes on to recount a story of a man in Georgia who was killed because of more than seventy stings from an Africanized hive. "He ran over the hive with his tractor and bumped it. And of course the bees are going to defend and get angry. And that is a lot of stings." He later explained how hysteria about Africanized bees has been generated to increase fear, thus establishing reliance on industrial solutions to bees.

Kneading some propolis (a Silly Putty–like substance produced by bees from tree resin) from the hive, Comfort continues: "The Africanized bee story became big because this huge lobbying point for the extermination industry—which is bigger than the bee industry. So now an exterminator can go to the bee call in Florida at some house and say, 'Oh they are Africanized—they are killers and I am going to have to charge you three times as much.' You have to use these extra products. And all these researchers get all this extra money to study these bees because the public is so scared of all these killer bees. This whole position is getting funded by this fear-based garbage." When pressed for his own interpretations of how and why Africanized bees might have been miscast, Comfort suggests, "There is a total stereotype. And they are from South America so we need more border patrol. The Africanized bee story has been part of this century-old war against wild bees saying that they are these cesspools of disease. Or these bees are surviving and our domestic bees are dying so we must treat them. Keep treating them. Whereas these wild bees are surviving and we don't want them because they are Africanized bees."

The way beekeepers talk about and manage bee populations also reflects cultural assumptions about the relationships among race, ethnicity, and nature. According to our fieldwork, the queen should steer clear of "feral" or "wild" drones who can potentially sully the metaphorical bee bloodline. However, in some cases, alternative beekeepers praise the feral or Africanized bee lines as somehow more primitive and unsullied by culture. Naturalizing arguments have also been employed in an attempt to delineate difference between human "races." From 19th-century "craniometry" (a science devoted to comparing the cranial capacities of humans), to the controversial book *The Bell Curve*, racist arguments have long been rooted in the natural body. What can be made of this ethnic/racialized discourse on honeybees and their human managers? We suggest that as humans we employ bees to buttress human racial boundaries. Africanized bees, for example, become discursively produced as already embedded with human fears about racial differences. This slippage between racial purity and fecundity of certain groups of bees to that of certain groups of humans is uncanny. In the early 1900s, a time of great immigration and internal migration in the United States, Theodore Roosevelt used the term "race suicide" to refer to the anxieties about the slowing reproduction of second- and third-generation Anglo Saxons.[35] Simultaneously with the decline in the white birth rate, the fertility rate of immigrants and minorities was rising. These themes of pronatalism and eugenics re-emerge during World War II.

Yet, unlike humans, bees are perhaps more "scary" as (documented and undocumented) immigrants because they come from another continent; fly across borders without permission for entry; have different norms, values, and taboos; and are assertive and reproduce with no regard for racial purity. Bees are not the only insect that has been linked to human racial difference, domination, and political boundary making. In Nazi Germany, Jews not only were likened to animals such as dogs and vermin but also were conceptualized as parasites, an effective biological metaphor used to separate unlike and unworthy species. In *Insectopedia*, the anthropologist Hugh Raffles powerfully portrays how direct linkages were made in the past by Nazis between parasites (lice) and Jews. He reprints the following language: "Lice are parasites (as are Jews.) They suck our blood (as do Jews). They carry disease (as do Jews). They enter our most intimate parts (as do Jews). They cause us harm without every knowing it (as do Jews). They are everywhere (as our Jews). They are disgusting. There is no reason they should live."[36]

With this appalling description, Jews, conceptualized as lice, are subject to the same ethical indifference that humans apply to many insects. Dehumanizing Jews and stripping them of their humanness makes them vulnerable to extermination through modern technologies. Human extermination and insect extermination are targeted toward organisms that are determined not only to have no human value but also to threaten the health of individuals and the state.

Africanized bees, or "killer bees," are insects that have been caught within a complex political landscape. Not only is the name "Africanized bee" itself a human construct, but the species has also been separated by its perceived "killer" behaviors thanks in part to cinematic portrayals and media coverage. It is no wonder that so many people, whether beekeepers, USDA officials, or the lay public, are concerned about what this species might do to us and our more docile (presumably) native bee populations. There is consensus that these bees are atypical, heartier, and more aggressive than species we are familiar with. Often African bees are described simply as "mean," according to the naturalist beekeeper Kirk Anderson. In a lecture he responded to the question "What do you think about mean Africanized bees" by saying, "There are different reasons why bees are mean. They could be sick or missing a queen. You would think that there were never mean bees ever . . . when they are mean people just say they are 'Africanized.'" In Anderson's experience, any bees that do not behave correctly or act assertively get labeled as Africanized. This is tangible proof of ethnic/racial stereotyping as applied to insects.

Regardless of how we racialize them, these bees are migrating farther north each year in the United States, adapting to local environments and climates. Humans rightly fear these bees because they are autonomous and seemingly out of control—they may potentially disrupt our "natural" way of life. So-called Africanized bees do unequivocally act differently when compared with other species, but this difference is contingent upon human discourses. And as some beekeepers argue, there are other frameworks in which to study and consider the characteristics of Africanized bees. Anderson asserts that he knew this: "They are strong. They are survivors, and they are not sickly." In his words they are "good hearty bees" that need to be seen and appreciated outside of the pejorative traits that humans automatically apply to them. But because race and ethnicity are so rooted in biology and genetics, Africanized bees will likely not be liberated from the political and cultural categories they are naturally defined by.

In addition to the racialized taxonomies and metaphors that seem commonplace among many beekeepers, there are greater logics at work in regulating bee immigration, migration, and reproduction. Similar to other animals and plants, such as demonstrated by the anthropologist Sarah Franklin's analysis of sheep in England, and the anthropologist Cori Hayden's discussion of ethnobotany in Mexico, honeybees are policed and politicized as "members" of nation-states.[37] Animals and plants become totems of national identities and are woven into the cultural ethos of human life. These totemic representations (images of woolly sheep on English hills and valleys) interact with the actual animal (the sheep) to recreate policies and sciences about how to "improve upon" the animal through the cloning of Dolly the Sheep. Our relationships to animals are not simply relationships of instrumental utility—honeybees make honey and pollinate our flowers, and they are also relationships about our feelings and beliefs about what animals symbolize. European honeybees are good immigrants who over time have become good citizens, working hard and building better lives for themselves and the rest of us. Conversely, Africanized honeybees are dangerous and sneaky border crossers who aren't properly socialized to our domestic ways. We can't easily add them to the melting pot.

In New York City, origin stories are told about European honeybees. These honeybees, Italian, Russian, Carniolans, have a classic historical legend as the wanted, desired, needed, and preferred immigrants. They were welcomed into our interspecies "melting pot." Honeybees are production animals, or animals with a primarily economic function to their keepers, and their stories of migration cleave with U.S. migrant labor stories.[38] However, as threats of illness, disease, and pests terrorized the assimilated population of honeybees (those who became "European" through their migration), borders were closed and only temporarily opened now and again and after careful scrutiny from human scientific interlocutors. This narrative of sanctioned immigration directly opposes as well as amplifies the threat of the stealthy and illegal border crossing of Africanized bees through our southern borders.

Becoming American Bees

One of the central questions the political theorist Jacqueline Stevens asks is how human groups are constituted. We have worked to understand how the forming of human groups extends well beyond our species. We also examine the traffic between the establishment of human and nonhuman species by governments and human/nonhuman animal actors. Through border patrol management, eugenic reproductive policies and practices, and miscegenation laws, human beings have attempted to construct a nationalist body that includes all the desirables and contains others in bare life (bare life is the life of a human stripped from any rights, living as animals with no political freedom): present, useful, and used but not members of the body politic. As Stevens further explains, "Every political society bases rules of inclusion and exclusion on invocations of birth, and every political society has a sex/gender system that exists via its kinship rules."[39]

How humans talk about and regulate bees can also be used, expanded, and perhaps reconceived through considering relationships between humans and nonhumans. It is not just humans who are labeled as "deviant," "outlaw," or "alien" by the state through ceremonies and rituals of their birth—nonhuman species, including insects, are also made legible—through dichotomous distinctions like good and bad, healthy or sick, docile or dangerous. Arrangements of selective breeding, hybridization, and miscegenation, as well as regulations over immigration, "entail racism," as Stevens argues about humans' constructions of political society, family, and nation.[40] Scientists have described bees as a "eusocial" insect species.[41] Eusociality implies a high level of social organization that involves intergenerational members, specialized divisions of labor, and coordinated care for the young. In other words, bees have an extreme form of kin selection—they choose their queen, accept her, and swarm when they decide it is necessary for the colony to split.

Insects and animals are thought to be easy targets as sites of control because of humanist ideas about who controls the planet—humans presume that we have a duty and a right to keep our indigenous populations free of unwanted species. This presumption of our duties and rights regarding nonhuman animals connects to our theme of a divine right over insects. Our presumptions are unchallenged and presented as natural and logical—for the greater good of a nation's citizenry and animal population. Importantly this story of insect immigration and migration occurs alongside that of humans' migration.

In the contemporary post-9/11 moment there is a proliferation of laws and regulations that assiduously attempts to maintain national boundaries even in light of the complexity of human migration and traffic of consumer goods and services. For human populations, Homeland Security now supervises immigration. Nicholas De Genova, an anthropologist, offers this definition: "Migrant 'illegality,' however like citizenship itself, is a juridical status. It signifies a social relation to the state; as such, migrant "illegality" entails the production of a preeminently *political* identity . . . within the regime of US immigration and naturalization law, it is noteworthy that the term 'immigrant' is reserved only for those so-called 'legal' migrants who have been certified as such by the state."[42] Like humans, bees have political identities in their status as legal or "illegal" migrants. De Genova goes on to explain the creation of "*deportable alien*" as a new status that produces a new population of human migrants that can be divided and repurposed for any activity of the state. Clearly, it has not been possible to deport bees as a species, but labeling them as a menacing pest that threatens the sanctity of the hive creates discursive and material conditions, specifically related to Africanized bees, that reinforce the illusion of border patrol and security.

Humans suffer at these borders, particularly Mexicans who endure harsh environmental conditions and mistreatment and violence, but do bees suffer from the regulation of national borders and boundaries? This question might rankle some. In sharing our book with many others, it is striking to us how many readers felt that it was "offensive" and "politically and professionally dangerous" for us to compare the suffering of an ethnic group of people to a race of bees. However, we feel this type of interspecies comparison of abject oppression and population displacement can enable us to better understand the ways we dehumanize groups of people by treating them like animals and how we disregard animals has having any "rights."

Becoming American is a dangerous proposition for both bees and humans. You must behave in the appropriate customs of the dominant group with respect to the norms of gender, sex, sexuality and reproduction, the consumption of resources, and the contribution to the economy. And if you are not born here, you also must enter through the sanctioned ways. Africanized bees are singled out for extermination in many different regions of the United States. In fact, there are only two countries that Americans can import bees from. We recognize that this extermination occurs because of the fear, the imagined and experienced threat, of "different" bees inhabiting domestic human-bee spaces. How can we understand this as suffering? And how should we recognize this insect's suffering? We are aware that these comparisons between human and insect suffering can trigger human responses of incredulity and frustration. As the sociologist Arnold Arluke queries, is it possible that a nonhuman creature "tarnishes or cheapens whatever group they champion and somehow, in their minds, trivializes the very notion of oppression"?[43] We stand firm that this insect suffering matters and does not deprive humans of attention and activism to their own suffering. Indeed by considering the ways humans regulate bee populations, the swarms, migrant experiences, and border crossings, we can perhaps better understand and eradicate suffering among all species, human and nonhuman alike.

7

Deploying Bees

The Work of Busy Bees

Arriving at Rockaway Beach in Queens at precisely the same time, we waved and exchanged beaming smiles. Quickening our steps along the boardwalk, we are filled with the excitement of urban researchers out of their normal gritty habitat. We are on a break from the "field" of rooftops, postage-stamp-sized backyards, cement-enclosed areas. The beach, misty with shore breezes and replete with wet-suited surfers and metal detector–wielding beachcombers, does not generally bring to mind bees and beekeeping. But on this glistening mid-September morning, we were attending the First Annual New York City Honey Fest. Similarly arranged to an urban farmers' market, this fair was different in that each of the fourteen tables—save for the pickle vendor—was dedicated to all things apiary (see figure 7.1). Beekeepers scurried around, setting up their booths and ware; others handed out leaflets with the day's programmed events. including beekeeping demonstrations, honey extraction lessons, small business advice, policy updates, cooking exhibitions, children's mural painting, and a honey tasting contest. As the event kicked off, bees were also in attendance at observation hives that were already attracting a curious crowd of humans.

As with many street festivals, there was a palpable buzz as event organizers and beekeepers eagerly arranged the boardwalk and greeted passersby. Yellow and black signs and streamers welcomed people. We overheard questions like "How much honey does a bee make?" and "How many times have you gotten stung?" Soon lines formed at stations to do art bee activities, watch bees, or taste honey.

Children pointed out individual bees to their parents, narrating the movements of bees through the comb. Some renegade individual bees also made their way to the event, perhaps drawn to the honey displayed for public consumption. Several human participants were also dressed

Figure 7.1. Close-up of honeybees on a fresh comb. (Photo credit: Mary Kosut)

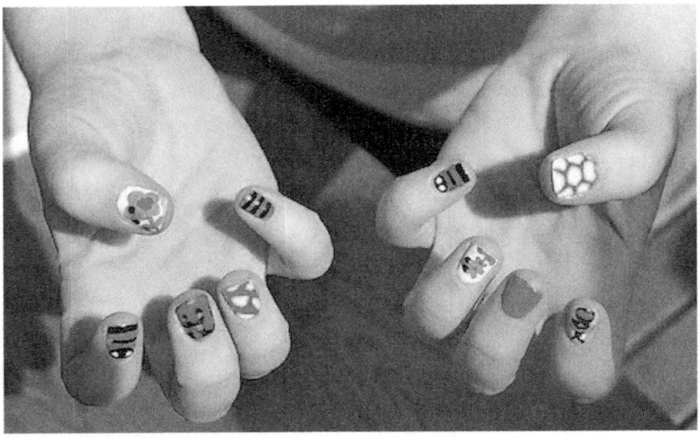

Figure 7.2. Bee-themed nails at the Honey Fest.
(Photo credit: Lisa Jean Moore)

for the festivities in bee costumes and the requisite yellow and black, accessorized with beanie hats and bee-themed cosmetics and jewelry.

A few of our informants were present and cheerfully greeted us, enthusiastic to share the celebratory aspects of the day (see figure 7.2). Backwards Beekeeping, in particular, had a strong presence at the event, with chief leaders in the loosely organized group performing solo and collaborative demonstrations (see figure 7.3).

Tim O'Neal had set up his extractor in the empty concessions area, and he brought in small groups of people at a time to witness the centrifuging of uncapped honey in the large circular spinning machine (see figure 7.4). Using a tool that removes the wax tops off of the framed honeycomb, the frames are then placed inside the extractor and hand cranked. Individuals lined up to try their hand at grinding the crank and we all listened to the whirl of the metal canister, imagining the gooey honey dripping down the sides of the machine. Once opened, we could see the sticky substance approaching the funneled bottom to be collected. You have to stop yourself from sticking a finger in to steal a little bit of the just-harvested treat.

Commerce was also part of the day with bee-themed silkscreened t-shirts and bags, local honey ranging from $7 to $15 per container, and beeswax candles. Honey-themed recipes were also displayed; beekeepers are eager to convert people away from the "evils" of sugar to the

Figure 7.3. Backwards Beekeeping table at the Honey Fest. (Photo credit: Lisa Jean Moore)

Figure 7.4. Honey extraction with full frames. (Photo credit: Lisa Jean Moore)

nutritional capacities of honey. Lists of the health benefits of honey consumption are passed out on multicolored glossy cardstock—say no to allergies, lose weight, live longer. Deemed a resounding success, the first annual Honey Fest, with more than one thousand people in attendance, sold out all its honey. A mixture of ecological activism, urban renewal, and locavore commerce, the Honey Fest demonstrates the overlapping interconnectedness of species and ecosystems—this is how bees help to bring humans together.

The Honey Fest

Throughout the year, New York City's public spaces are transformed into festivals mostly associated with ethnic celebrations—San Gennaro in Little Italy, the Saint Patrick's Day and Puerto Rican Day parades down Fifth Avenue, and West Indian Carnival on Eastern Parkway in Brooklyn. These festivals bring value to the city as they capitalize on ethnic and social diversity and enable "free" commercial zones to sprout up along the "festival" route. Honey Fest is unique as far as these free, publicly accessible, urban festivals go. Allegedly about celebrating all things "honeybee," this event is an interspecies collaboration between honeybees and humans. Honey Fest is slightly similar to the use of urban space for other nonhuman species, such as when FIDO (Fellowship in the Interest of Dogs and Their Owners) offers monthly human and dog treats in Prospect Park, creating opportunities for a congregation of canine enthusiasts. However, there are no commerce transactions occurring at FIDO events and individual dogs are fully incorporated as they mill around the dog treat table in various degrees of communication with their companion humans. We start from the Honey Fest because it illustrates that the commercial investment in bees is distinct from that of most other insects. While it is true that many natural history museums and botanic gardens offer butterfly tents for patrons to mingle among the butterflies, the production of commodities from the butterflies themselves has not created cottage industries.

Most people are not aware of the magnitude of our dependence on bees. Producing honey and pollinating is something the bees "do naturally," as part of their animal instinct, and it happens in places out of view for most people. Also as much feminist scholarship cogently explains, arguments about "animal instincts" are used to explain the undertaking of tasks like breastfeeding and caregiving—the lioness nurses her young and so did your mommy; it's natural. Creating this natural explanation makes invisible the actual social mandates about gendered divisions of labor. Once a human's or an animal's work is naturalized through explaining it as a mammalian impulse, rather than a socially regulated and mandated practice, the production is undervalued in the political economy. In the case of the bees, they naturally produce honey and therefore we need not acknowledge or compensate the labor that goes into the production. For humans harvest honey, and this work of extraction makes the commodity real to us. There is a particular buzz centered around events like Honey Fest—the buzz of human production and consumption. Honey as a raw material is transformed by human labor so that it has use value not only for bees but also for us. In the process of removing honey from the hive and extracting it from the comb, and then placing it in decorative jars and containers, it moves further away from the colony literally and symbolically. Honey is no longer food for bees to survive on over winter, but a product. And humans certainly get excited about this "natural" product as they buzz around, tasting, buying, creating, and socializing. Yet amidst all this buzz over honey, the bees themselves seem to get lost as humans take credit for cultivating or purchasing honey, especially under the umbrella of greenness or holistic health. Bees produce honey for *their* survival. We clearly want their honey but do bees "naturally" want us to have it? Honey Fest provides us with one way to witness how humans have used the bee for larger social purposes—from community building to urban renewal and holistic health practices to recipe concoction. Here we investigate the cultural traffic in honey between the local and the global. We acknowledge the friction between caretaking and exploiting of bees enacted by both lay beekeepers as well as corporate, commercial, medical, and military interests.

Honey Fest is just one example of how honeybees are enmeshed in hyper local cultures and micro-ecologies in addition to global capitalism and environmental politics. They fit in a larger political agenda as part of the sustainability craze and are often co-opted as part of corporate campaigns. In this respect, bees have become a useful and sometimes profitable "cause." Bee products are also revered for their medicinal benefits, particularly within the context of alternative health practices and increasingly within big pharma's pursuit for new treatments. In these instances, humans rally for the bee as bee health is interpreted through what is good for human bodies, as well as human habitats.

We consider this question cautiously and with a great deal of ambivalence. Perhaps, rather than speaking on their behalf, as some of our informants attempt to do, our project has been to examine how various people engage with bees. We continue to struggle with our own sympathetic or empathic confusion about whose story we are "telling" and whose perspective we are taking—the bees or the humans. Some humans see bees as vulnerable creatures to be cared for in the face of habitat destruction (enacted primarily by humans). But beyond the "stewardship" or "social action" of humans, there are several beneficiaries from the golden harvest of bees' productivity. Some urban beekeepers may be atoning for the sins of their species while simultaneously perpetrating a human-motivated violence against the bees. This violence happens to the bees in many contexts. Honeybees are diesel-carted, migratory day laborers for monocrops; their habitats are destroyed by human construction and climate change; and their bodies are sacrificed through accidental and purposeful apiary practices of hive checks, virus and mite treatments, and honey harvesting. The similarities between the treatment of nonhuman animals and human beings with respect to misuse and abuse have been discussed by many scholars—in particular, by examining connections among animal abuse and human slavery and the Holocaust.[1] We find these connections compelling as we imagine colonies of bees contained in semitrailers without air, food, or water in a state of semistarvation, primed to work the next crop.

Industrial beekeeping, which emerged after World War I due to the increase in highways and the growth of trucking, is quite different than the beekeeping practices we have studied.[2] Essentially up to 450 hives are loaded onto semitrucks and transported across the country to farms that "rent" the hives for a period of time. The industrial model of beekeeping uses the Langstroth hive, which is the typically white, stacked box made of wood most people are familiar with. In the words of a former industrial beekeeper, "When beekeeping was becoming a profitable business and needed to be standardized for suppliers to create monopoly," the Langstroth's place in agri-beekeeping was cemented. Suddenly, "the box hive was in every corner of the world, " because it is efficient, standardized, and convenient for humans to work with. While some question whether they are good for bees, the industrial hives are undoubtedly good for business. Many crops rely on bees' pollination for their continued propagation, including blueberries, cranberries, almonds, alfalfa, sunflowers, and apples—and rental rates for bee colonies can vary between $10 and $180 per colony, depending on the season and amount of time in use.[3]

As ethnographers, we are deeply invested in these relations of power and do not want to take our subjective position as researchers for granted.[4] There have been many stories told about bees recently in the press and by academics. Most of these narratives speak on behalf of the bees, but they come from different expert positions, including entomologists and beekeepers. All of the humans that surround bees get something from them, such as food, honey, companionship, a research grant, a book contract, and notoriety. So while we are all not extracting honey, or making beeswax candles, humans benefit from bees' labor every day of our lives. What are the different ways in which people come to *know* bees? We are aware that we ourselves will know bees only through the way other humans interpret them, as well as from our own personal exchanges with bees. There are many layers of humanness that mediate bees as simultaneous objects or subjects of research. Bees will never be our interlocutors, but that doesn't let us off the hook; we are not ethically exempt or free to exploit them in our narrative as research subjects. In addition to engaging in reflexive practice, in this chapter, we mine various locations where humans use bees as an exploration into the deep interconnections between our species.

"Stealing Honey"

People from many different walks of life—chefs, poets, business executives, entomologists, and feminists—remark on the industry of bees. Most commonly, we associate bees with honey, the amber golden liquid that sweetens tea, Cheerios, and beer. Honey is a natural substance that fuels folklore and philosophy, signifying sweetness and also persuasion, seduction, and the human soul. For example, the patron saint of honeybees and beekeepers, Saint Ambrose, apparently quite a miraculous speaker who proselytized many in the 370s, has been identified as the "honey-tongued doctor" because his words could sting. The folklore around Saint Ambrose holds that when he was an infant, bees swarmed over him in the crib. Instead of stinging the baby, they dropped some honey on his tongue, giving him the gift of oratory.[5]

In the 1920s, the Austrian philosopher and naturalist beekeeper Rudolf Steiner advocated for the use of honey as nourishment for the body and soul. Honey not only is delicious but also is transformative:

> Honey creates sensual pleasure, at the most, on the tongue. At the moment you eat honey, it creates the proper connection and relationship between the airy and fluid elements in the human being. There is nothing better for a human being than to add a little honey in the right quantity of food. In a very wonderful way, the bees see to it that a person learns to work on the internal organs by means of this soul element. By means of the honey, the bee colony returns to humans the amount of effort the soul needs to expand in their bodies.[6]

But more than making honey, these creatures also create beeswax and collect pollen. Beeswax is produced by workers (at twelve to eighteen days old) through their four wax glands under their abdomen. When a field or forager bee, legs loaded with pollen, returns to the hive, the house bees work to unload the pollen. These house bees then mix bee saliva, the plant pollen, and nectar. The secreted enzymes mix with the pollen and make bee bread. From tree resin, bees make propolis, also called bee glue, a resinous sap the bees collect and mix with wax flakes secreted from glands under the bees' abdomens. They fabricate royal jelly made by nurse bees through their hypopharyngeal glands. Royal jelly is highly nutritious and milky white in consistency and color. This food was originally made for queens, which is attributed to providing the queen's longevity, fertility, and size. Bees also sting with venom, or apitoxin. Humans have thought of ways to collect, use, and enhance bees' creations. Screens wipe pollen from bees' legs upon entering the hive, hive tools can scrape propolis from the nooks and crannies of the bees' home, and extractors spin out honey from frames. *Bee Culture*, a magazine for beekeepers, is filled with advertisements for devices that allow us to more efficiently extract what we want. These tools are designed to allow humans to get the most out of bees, in this case mainly honey for personal use or for profit.

Much human-bee interaction is based around honey production and extraction. The meaning of honey has shifted within the contemporary locavore zeitgeist and now holds culinary cachet. Honey, previously a curative to add to teas or cough drops for sore throats, or an alternative to sugar, has now become something to celebrate as a specialty ingredient worthy of a sommelier. Beekeepers proudly describe the unique characteristics of their honey, particularly the shade of color (ranging from light yellow to rich reddish brown), the smell, and the taste—honey has levels of sweetness that a discerning palate can identify. Honey, like wine or varieties of heirloom tomatoes, is increasingly utilized not only as a generic ingredient but also as a gourmet item. Gourmands and self-proclaimed foodies are eager to share their honey knowledge.

But bees "give" us (or we take) other pieces of their lives and bodies in the interest of somehow improving our own. Throughout our fieldwork and research, we were continually reminded of the utility of bees as well as the sundry ways in which humans have found work for them (or have presumed that what bees do "naturally" is raw material that humans are entitled to). Table 7.1 catalogs bees' products as well as some of the urban beekeepers' uses of these raw materials. This is not meant to be exhaustive; this table indicates what informants shared with us as their uses of api products and what we witnessed during our years of fieldwork. It illustrates the similarities and differences between how bees use what they make and how humans make use of every aspect of bees' lives possible. For example, bee venom has a different utility for bees than for humans. Bees manufacture it as a protective mechanism to be used to potentially spare the hive from harm, while some humans use venom to treat arthritis. Notwithstanding, honey, as a food source, is the only substance that fulfills the same need in bees and humans.

Table 7.1. Bees' Material Production

Bee Labor	Use by Bees	Use by Humans	Method of Extraction by Urban Beekeepers	Method of Extraction for Industrial Beekeeping (500 or More Colonies)
Beeswax	Building honeycomb cells as chambers for brood and for storing honey and pollen	Candles, cosmetics, soap making, furniture polish, waterproofing material	Taken from the hive through disassembling frames—particularly when hives "crash" and melt down	Similar process on a larger scale
Honey	Food source	Food source	This is the most common self-reported reason that urban beekeepers maintain hives; removable frames and extractors	Most of the honey consumed in the United States is not produced domestically
Pollen/ Bee Bread	Protein source for feeding brood	Dietary supplements, treatment of allergies	Placing a screen at the opening to hives to scrape off pollen	Generally not a product among commercial beekeepers
Propolis	Sealing the hive	Medicinal uses for coughs, inflammation, allergies, cuts, skin conditions	Scraped from boxes with hive tool; mixed with alcohol in tinctures	Generally not a product among commercial beekeepers
Royal jelly	Growing queen bees	Medicinal uses for many medical conditions as well as immune system boosting	Only one beekeeper has extracted royal jelly by harvesting it from queen cells	Generally not a product among commercial beekeepers
Venom	Defensive use through stinging—resulting in the honeybee's death	Medicinal use for bee-sting therapy	Beekeepers report purposely stinging themselves in fingers, hands, and knees	Generally not a product among commercial beekeepers
Pollination	Collection for food	Fertilization of crops	More prolific gardens are cited as a secondary benefit among urban beekeepers	Primary reason for U.S. industrial beekeeping is migratory pollination: bees are rented out, trucked across vast terrain, and directed to pollinate monocrops, which are overwhelmingly almonds
Bees	Reproducing the hive	Starting a new colony of bees; requeening a hive that is queenless	Raising bees to split the hive and send off a nucleus hive; queen breeding	Queen breeding

With the increase in urban beekeeping in New York City, we found that most urban beekeepers harvested and either sold or gifted their honey. Honey becomes empirical evidence of a job well done and is a reward for being a skillful beekeeper. Of course, the bees don't make it as a gift, but beekeepers are excited to "receive it" nonetheless. Our informants often mentioned how much honey they had harvested in previous seasons, ranging from 42 to 350 pounds. As neophytes ourselves, it was difficult for us to gauge how to react to these numbers particularly because each beekeeper would react differently to different amounts—"We harvested forty-two pounds last year so that was great" or "I got one hundred pounds and that was disappointing. I had hoped for more—there is always this [referring to the interview year] year." But despite these varied evaluations of the size of harvests, most urban beekeepers use their proceeds to reinvest in their hobby by purchasing new equipment or packages of bees for the subsequent season. As described below by B.J., gifting honey is also a popular practice that is deeply appreciated:

> I sell my honey in a minute and mostly to my colleagues. I don't do farmer's markets or go to stores and I have yet to design a label. I am not really a businessperson. I feel like a little cottage industry. I make a profit and the money I use to do things like buying an extractor. I love this backyard and doing this. I love that my colleagues are so grateful and tell me it is the best honey I ever had. Your honey tastes like sunshine. Spring to fall I can see the difference. The fall is much heavier—a much heavier flavor. Spring honey is very light and lovely. The texture is basically the same but the intensity is different. I have also given my honey as gifts. One time I remember really clearly. I am friends with a very important opera star and I catered a dinner at her house and I brought her a jar of honey and she said, "You know what, no one has ever brought me a gift like that and that actually meant a lot to me" because I am sure she is showered with gifts all the time. If I go to someone's house I bring them eggs or honey or candles or bread or yogurt. I think that is something that is meaningful.

Another beekeeper explained to us how she designed her own labels and created a website for advertising her honey as well as advocating for locally produced honey. Through her connections, she developed a relationship to Murray's Cheese Shop in Manhattan, where she was able to quickly and consistently sell out her jars with a demand for more each season. Murray's specializes in gourmet and artisanal cheese and obscure items; it is legendary within New York foodie circles. True to another trend among the informants, there is a seemingly unquenchable demand for local urban honey—indeed, many called it "liquid gold" and indicated that the more honey they could harvest, the more that they could sell. The market is also possibly expanding within locavore consciousness as news of Chinese honey laundering becomes more well known (for example, there was a recent story on National Public Radio, as we explained in chapter 3). One of the nation's top honey packers has said, "With all the food safety and food security issues, knowing where your food comes from right now is incredibly important."[7]

At New York City farmers' markets there are a number of different tables dedicated to the sale of honey and bee-related products, and beekeepers advertise with labels that authenticate Harlem, Red Hook, or Greenpoint honey. Buying honey from a particular borough demonstrates the sociologist Thorstein Veblen's theory of conspicuous consumption, illustrating a symbolic advantage that one group of people has over others.[8] As he explained, the public display of luxury items—the things we want, not things we need—indicates the purchaser's social status for others to witness. Buying expensive and cottage-produced honey at a farmers' market, publicly demonstrates the purchaser's sense of taste, knowledge of food, and financial stability. Those who shop at specialty boutiques, gourmet shops, or farmers' markets, can afford to care about the local greening of New York City and can pay for honey hailing from microregional ecologies. Gourmet local honey may cost twice as much as the generic plastic honey bears sold in chain grocery stores; generic honey can run from between $4 to $7 a jar, whereas comparably sized gourmet honey sells between $8 and $14. However, honey, while consistently produced by bees through spring, summer, and fall, is not an endlessly produced or unlimited resource in the Northeast. In New York City, humans are in direct competition with bees for honey. In order to increase their chances for survival, bees must have resources or they will starve to death over the winter. Striking the balance between how much honey stores to leave for the bees over winter is debated by urban beekeepers. As noted by one beekeeper, "I think my first hives were lost because I just didn't know what I was doing and they died of starvation. So now I am sure to leave a lot of honey frames in with each hive."

"Stealing from the bees," as it is sometimes called, indicates that some humans are aware that harvesting honey must be managed in light of the bees' own needs. Empathetic beekeepers, those who spoke of being responsible stewards, talked to us about what could be called an ethics of honey harvesting. But it is humans who ultimately decided how much honey fulfills the needs of the hive. In this context, honey is at the nexus between what bees require for survival and what humans want. Honey is the primary food source of a hive; its absence or presence means life or death. Humans can survive without honey, which means that we have a different relationship to this substance. Its human use value runs the gamut from pleasure, to dietary preference, and, in some cases, to economics. When humans say that they are working to keep their hives alive and trying to give (or leave) the bees what they need, self-interest may corrupt and cloud the intentions of even the most attentive beekeepers. Indeed several beekeepers told stories of their earlier careers when their bees either froze or starved to death because of improper preparation for winter and overzealous honey harvesting.

For most urban beekeepers, their practice is a hobby or activity that puts them into social situations with others who are dedicated to local urban agriculture or urban homesteading. Maintaining one's livelihood exclusively based on beekeeping is decidedly more difficult. Our fieldwork included only three members of the urban beekeeping community who were able to earn a livelihood from their practice. Each of these three beekeepers, however, did so through multiple practices with bees, teaching, honey sales, pollen sales, queen grafting, writing about bees, and public speaking—as exemplified by Meg Paska's comments:

> I earn most of my living as a beekeeper. I write articles for *Bee Culture* magazine—a monthly column. I just got a book deal on how-to urban beekeeping, with recipes, in early 2013 from Chronicle Books. [World-renowned chefs] Mario Batali and Joe Bastianoic pay me to keep bees for them at the Terry Lodge in Portchester. I also teach classes at Third Ward

> [an adult learning center in Brooklyn]—a lot of members and it is a lot of young people—late twenties and early thirties, varied, a lot of women. It's way more empowering. I want to lay the groundwork now: I want to find a way to give back. I want to find an organization that focuses on assisting women from domestic abuse or sexual assault and teach them how to keep bees as an empowerment exercise and to generate some income for them and their families—that and my bee yard is what I want to do next. Their honey is the primary thing you start harvesting and then you can move on once you get more comfortable.

True to her social activism commitments, Paska explains how beekeeping can be taught to others as a form of social and economic transformation. It is reminiscent of the Taoist proverb "Give a man a fish and you feed him for a day; teach a man to fish and you feed him for a lifetime." Teaching a woman how to be a beekeeper could generate some income to transform her economic circumstances.

As part of this interview, we asked Paska if she thought the bees realized that she was taking the honey and how she felt about that.

> Bees might know you take the honey; they might recognize at some point that you are taking from them, but they are so busy that it is like, on to more pressing matters. They have memory in terms of navigation. They recognize human faces as a pattern of facial features that differentiate us from other creatures. I think bees just get accustomed to being handled and a certain style. That is how far I think they know me.

According to a team of international scientists, Paska is accurate in her claim that bees can recognize human faces as distinct from other faces—but it is not proven that bees can distinguish between individuals.[9]

As bees are housekeeping, raising brood, queen grooming, foraging, and navigating, they might not have time to contemplate the human theft of honey. As Paska suggests, bees have learned to work with us because it is expedient for their own survival. Bees adapt to humans, and unless they sting us, it is on the humans' terms. There is interspecies socialization or, as Tim Morton calls it, "the mesh," whereby all forms of life are inextricably entangled within an interdependency that erases any culture and nature distinctions.[10]

Rather than accept the notion that humans help save the bees, we argue that humans raise bees for very particular purposes that for the most part do not directly benefit the bees. Furthermore, it is difficult to receive feedback about whether bees, as insects, are "happy" or agreeable with what we are doing than, say, we would get from a dog wagging its tail. The bees can tell beekeepers to get away by stinging, but we assume they will be able to adapt to human intervention and interference. We are suggesting that people might not be able to determine "what the bees want" and rather are pushing what bees will tolerate. We, as humans, adapt as a species to new systems of rationalization and capital extraction. To continue to compete in late capitalism, we must be able to multitask and stretch our own flexibility. Because we are so deeply interconnected with bees, we expect them to be able to adjust and adapt to new circumstances of labor just as quickly as we are required to do.

The Bee as Brand

Increasingly, we buy into bees as an ecocultural symbol. In a follow-that-bee approach, we chase the insect through green consumerism (by means of the ecological movement). *Save the bee* t-shirts and honey-based products often share an ideological connection to so-called alternative lifestyle practices. There is an ethics to being and living green in terms of saving the planet under the rubric of sustainability. But bees are also part of the business of being green. As a *New York Times* article states, "Some 35 million Americans regularly buy products that claim to be earth-friendly, according to one report, everything from organic beeswax lipstick from the west Zambian rain forest to Toyota Priuses."[11] The merger of green consumerism and ecofriendly politics both defines and drives the "current environmental movement as equal parts concern for the earth and for making a stylish statement." Given the aesthetically pleasing nature of bees, their medicinal uses, their crucial role in our food supply, and the emergence of Colony Collapse Disorder (CCD) and the subsequent media reportage on the subject, we argue that honeybees are being consumed by green consumerism itself.

Like a dog is "man's best friend," used to signify fidelity in children's books, family television programs, popular movies, and Western art history, certain bees signify wholesomeness, healthy living, hard work, productivity, and creativity. Clearly there are other more negative symbolic meanings of bees—as lethal or dangerous—but these meanings of the bee seem to fade when the creatures are aligned with specific products. Positive traits, layered upon the bee, are aspirational for humans. Through socialization, humans are ideologically encouraged to value certain ideals—hard work and industry, like a bee's, achieves material success and admiration. As with the construction of hegemonic messages, there is the erasure of the cultural work that humans perform in the process of naturalizing bees as the model insect, a desirable species with original and inborn traits.

Sociologists have suggested that in contemporary Western cultures, individuals no longer chiefly identify with their productive capacities as beings; rather, they have shifted their self-concept toward consumption patterns.[12] We are what we buy. People turn to their acquisition and relationships with their possessions, including animals and objects, as a part of their self-identity. "Who am I?" is often answered by what I own, where I shop, or what products I use. What we buy then becomes a reflection of who we are as individuals—which is counterintuitive in that we are staking a claim in our sense of being unique while at the same time aligning ourselves with a group of other consumers. The trick for any marketing campaign is to tap into those identification desires, the self we want to demonstrate to others, and to provide a product that enables the "real" us to be revealed. With the rise of sustainable or green consumerism, there is an ironic tension between the desire to express one's identity through consumption and the desire to signal an environmental conscious self who does not waste resources. And the marketplace has a solution for this tension in the consumptive practices that both sustain the environment as well as the socially and ecologically conscientious self. The sociologist Dennis Soron writes, "Appealing to this individualistic ethos, the consumer marketplace has increasingly presented people in affluent societies with opportunities for greening their identities and putting a version of 'sustainable consumption' into practice. Indeed, it has tossed up an abundance of conscience-soothing goods, ranging from benign items such as sustainably farmed produce and energy-efficient appliances to heavily processed and packaged organic foods and outright absurdities such as 'eco-friendly' cigarettes, bullets, plasma televisions, yachts and SUVs."[13]

It is tricky business both to appeal to people's desire to own things and to their desire to "live simply." Community-supported agriculture (CSA), a practice where a group of neighbors buy into a farm's harvest and receive produce throughout the year, has been successful at attracting people to participate in a marketplace that limits their choice (through having less produce available than supermarkets) and yet enables consumers to feel their "ethical consumption" is part of "enchanting moral virtues."[14] As the philosopher Clive Hamilton suggests, "The task of the advertising industry is to uncover the complex set of feelings that might be associated with particular products and to design marketing campaigns to appeal to those feelings."[15] One method for pursuing this task is to purposefully select a branding strategy that creates a personal identity for a product. Bees are the perfect solution—they are threatened, ecologically friendly, and perform the trick of absolving people of their need to own things. And so the bee is a go-to representative for diverse products, including honey, beer, sports teams' merchandise, tools, wireless services, salmon, tuna, and soap.

Burt's Bees, established in 1984, is a prime example of how bee culture has been visualized and commercialized (see figure 7.5). The company started as a beeswax candle offshoot of someone's hobbyist honey production. Over the next two decades, the company experienced remarkable success. In a 2006 interview with *Business Week* magazine about the best global brands, John Replogle, the CEO, stated, "Health and wellness [are part of] a megatrend, and so is the greening of American. All you have to do is listen to major retailers talk about sustainability, and you see the confluence of consumer trends and retailers' trends. We're right at the heart of that."[16] By 2007 Burt's Bees had grown into a $250-million company, with products available beyond health food stores at megastores like Target and Rite-Aid. The success of the company has not gone unnoticed by larger corporations. At the end of 2007, Burt's was purchased by the Clorox Corporation for $913 million. As stated by Clorox representatives, the bleach giant hopes to model the greening of its brand through this acquisition.[17]

True to its sustainability mission, Burt's Bees had a campaign to give 5 percent of proceeds for CCD research throughout 2008. Over

Figure 7.5. Burt's Bees cosmetics. (©2013 Burt's Bees, Inc. Reprinted with permission)

half of their products—which include lip balm, facial products, baby lotions, hair care, cough drops, toothpaste, and (somewhat ironically) bug repellent—are 100 percent natural, and many, although not all, products use ingredients from bees such as beeswax, honey, bee pollen extract, and royal jelly. Burt's Bees website states, "And of course, we will never conduct product or ingredient tests of any kind on animals. A few of our products contain ingredients derived from animals such as beeswax, royal jelly and milk. And for that, we thank them."[18] This acknowledgment of the bee works to appease human consumers of Burt's products, although bees and cows are probably unaware of our gratitude.

Regardless of animal and insect relationships to the company, the use of the bee both in the title of the company as well as on its product packaging has been consistent since its inception. The image of Burt is modeled off the actual Burt Shavitz, a Maine beekeeper who teamed up with a single mom named Roxanne Quimby to start the company. It is striking how the image of the bearded, rugged man is similar to

Grizzly Adams or some survivalist earthy mountain man; this imagery is in and of itself iconic of the "natural human." Like some Thoreau-esque signifier, Burt's identity is for sale—with bees buzzing around his head, embedding him within an ecology of rural tranquillity. By purchasing Burt's Bees products, even if you can't go be/live in nature, you can buy nature. You signal to others that you care about nature. You are a natural beauty, rather than an artificial one. Here, perhaps, the bee and the honeycomb, integrated into all marketing and product imagery, are symbols of Mother Nature. Humans are saving this mother by buying ethically and feeling connected (but through our credit cards rather than an actual walk in the woods). Bees, in their endless self-sacrificing toiling, have provided an opportunity for even a company aligned with household and industrial bleach, which contain known caustic agents to humans and the environment, to atone for its sins and for humans to have soothing contact with bees on their lips, skin, and hair. Here, bees can make us look and feel better about ourselves.

Apitherapy: Medical Uses of Bees

Health claims about the medicinal use of bee products cleave with these contemporary food and environmental movements. Some people conceptualize and employ bee food as part of a cleaner, healthier alternative lifestyle, as complements to their uses of yoga, meditation, and organic foods. Humans have come to the aid of bees when their populations were decimated by what is now clearly a panoply of causes, including illnesses and diseases. Ironically, though, bees and their products have been in the service of human health for centuries.

Among our informants, several individuals swear by the positive and beneficial effects of anything derived from the hive and its inhabitants. From soothing dry coughs to curing eczema, treating arthritis to fortifying fertility, a variety of apitherapies are touted as medical miracles.[19] The use of apitherapy, or the medicinal use of honeybee products for human ailments, has existed since the ancient Greek physician Hippocrates's use of bee venom and propolis in treatments.[20] The American Apitherapy Society, "devoted to educating about and promoting the use of honeybee products to further good health and to treat a variety of conditions and diseases,"[21] is a member-based organization that was founded in 1978. While it is not our intention here to verify that apitherapy is a safe and effective health practice, we both have used these therapies by eating bee pollen and taking royal jelly. Our purpose, first, is to examine how apitherapy is used in the everyday lives of beekeepers and others and, second, to understand what apitherapy might mean for human-insect relations.[22]

Traditional Chinese herbalists have acknowledged the beneficial effects of bee pollen and royal jelly for centuries, but it is only recently that the fruits of the hive, pollen in particular, have been touted widely as the "perfect food" or "complete food" in media and health-related discourses. This is because the pollen, residue from plants that bees bring back to the hive to be stored in the comb, is extraordinarily rich in vitamins, minerals, and protein. Pollen and honey have been used in ancient and non-Western health and healing practices, and they have gained currency in the late 20th century due in part to the recognition of alternative medicine. Thus, bees are situated amid a variety of holistic lifestyles and regimes and are also in the process of being co-opted and economically exploited for global biomedical and pharmaceutical enterprises.

Medihoney

In the case of medicinal uses of honey, not all honey is created equal. Manuka honey, a type of mono-floral honey (from only one pollination source), hails from New Zealand and is made by bees who visit the manuka bush, called the *Leptospermum scoparium*. The flowering of this plant lasts only six weeks each year, and bees pollinate it for roughly two to three weeks. The honey does travel widely, as it is available in Europe and the United States. This antibacterial honey has been clinically proven to shorten the median time of wound healing[23] and has passed the FDA's first round of an approval process. The most promising of the results is that this type of honey is effective against methicillin-resistant *Staphylococcus aureus* (MRSA, or multiresistant germs).[24] Investment and development into creating agricultural circumstances for more Manuka honey have increased because this honey source is anticipated to be a $1 billion industry for New Zealand's Ministry of Agriculture and Forestry in conjunction with Comvita (the parent company of Medihoney, which owns 75 percent of the Manuka honey supply).[25] However, there have been reports of a shortage of trained beekeepers in New Zealand,[26] prompting the government to actively recruit overseas beekeepers to consider relocating to their country.

Medihoney is the brand that produces wound-care dressings for human use, including gel sheets, tulle dressings, wound gel, barrier cream, and apinate dressings. These treatments are available in different strengths, both over the counter and with a prescription. Claims about medihoney treatments include that they are faster, cheaper, and more effective with limited side effects compared to conventional medicine. The tag line to the product is "It's not just any honey, it's MEDIHONEY," accompanied by an icon of three honeybee hexagons with a plus sign or a cross to indicate first aid. Used to treat human ailments such as leg and foot ulcers, pressure and diabetic ulcers, burns, surgical wounds, bedsores, and abrasions, New Zealand bees' honey has been harvested and modified to enable healing through biochemical processes favorable to cell regeneration. Humans have made honey better for humans by improving on the bees' own labor; the honey has been medically modified to heal fleshy, oozing human bodies. We would venture to say that, with the rising rates of chronic conditions such as diabetes in the United States with slow healing and infected wounds, it is likely that the demand for Manuka honey will keep growing.

It is nearly impossible to find any readily available information on how the bees are treated by Comvita in the Medihoney production process. Presumably few bees die in the harvesting of Manuka honey for Medihoney, and yet it is difficult to imagine how the bees are compensated for their labor. It strikes us that Medihoney, its innovation and use, illustrates another example of how bees make themselves indispensable to humans. While honey (raw or treated) has nutritional and economic value, a new modified and medicalized honey may lead to the increasing medicalization of bees.

Bee-Venom Therapy

Nonhuman animal-assisted therapy (AAT) is the practice of using animals to increase human mobility and accessibility as well as for various treatments for mental illnesses.[27] The American Veterinary Medical

Association defines AAT as "a goal directed intervention in which an animal meeting specific criteria is an integral part of the treatment process."[28] Most commonly, "dogs, cats, rodents, birds, reptiles, horses, monkeys and even dolphins" are employed as service animals.[29] Debates abound about the difference between use and exploitation of animals but most scholars agree that AAT has been heavily human-centered in its execution. Humans do go to great lengths to justify their practices with service animals—such as the animals "enjoy the practice" and like to be "useful" or that AAT practices create opportunities for animal integration into human lives and hence can increase the value of and improve the animals' own treatment. With respect to bees, this does not seem to be the case.

Notably absent, insects are not considered in the AAT literature. Among the health and healing practices associated with bees, we have chosen to examine Medihoney dressings and the naturopathic health practice that was gingerly attempted by our informants, called bee-venom theory (or bee-sting therapy). We were first introduced to the idea of bee-venom therapy (BVT) through our meetup group lectures. BVT is the practice of injecting bee venom through the stinger to alleviate inflammation and joint pain. Jim Fischer stated, "There are some people who claim that their MS was cured by bee stings. Something medicinal about the bees, and I don't think it is good for beekeepers to make these claims. But there is some evidence to suggest bee venom helps some people. But if bee venom had something worthwhile to offer, then big pharma would have figured it out." The expense of clinical trials and the unpredictability of bees as medical research subjects may have indeed hampered the use of bees in bee venom case-controlled studies. In fact, many of our informants have engaged in this practice of selecting bees from their hives to strategically sting them in specific parts of their bodies. This deliberate stinging is quite different than the bee's own defensive action of stinging as a last resort to protect the hive. Instead, the beekeeper is in control of the stinger and has to poke himself with the bee's abdomen in hopes that the stinger will be released and the venom will be pumped through the skin. The bee will die.

A honeybee's stinger is attached to its abdomen and is made of sharp barbed lancets and a venom sac; when the bee stings and flies away, the barbs stick into a person's body, pulling part of the abdomen from the bee's body and eventually killing the bee. As one informant described, "It is like having your whole lower half pulled off." There are several active proteins in bee venom: two significant substances are "melittin" (a powerful anti-inflammatory) and "adolapin" (an anti-inflammatory and pain-blocking ingredient). The challenge with bee-venom therapy is that a small segment of the human population (approximately 1 percent) will experience anaphylaxis, a life-threatening type of allergic reaction, to bee stings. Once it is certain that a human does not have this reaction, he or she could be eligible for BVT. Videos from YouTube demonstrate this extreme new health craze[30] whereby bees held by tweezers are repeatedly pressed up to human skin until they sting.

A backyard beekeeper in Cobble Heights, Brooklyn, explains, her experience with BVT: "Two years ago I had tendonitis in my thumb and I could not even pick up a plate. So I went to an apitherapy meeting and they were talking about sting therapy. So I did. So I had her sting me here and she detached herself. I thanked her as she died. And I took the stinger and stung myself twice more—gone. All gone." A stinger is still able to pump out venom once it is pulled out and used to rejab the skin. She flexed her hands and showed the dexterity of her thumb.

Another urban beekeeper, Christine, explained how she and her partner have tried to sting themselves as part of apitherapy for the arthritis in their knees, backs, and fingers. "At first we picked a drone, before we knew what we were doing and tried to get the drone to sting us and then we realized it didn't have a stinger. So then we got a worker, we wanted to pick one who was a forager and later in life. You just hold it by the wings and try and poke yourself with it to get it to sting in the right place, which is really harder than you would think." Christine moved her hands to show that her pinkies were gnarled. She added that she did feel slightly badly that the bee "is dying for my arthritis." But she was relieved that the use of venom appeared to be working at easing the symptoms of her condition.

Immediately after each interview, when we debriefed, invariably, our conversation quickly turned to recounting these practices of self-stinging. It seemed incongruous to us that beekeepers who so clearly wanted to protect, help, and care for bees, would, at the same time, see the bee's life as worth sacrificing to alleviate their own pain. Furthermore, some of the same beekeepers who had shared with us their emotional reactions to bee death participated in BVT. Had we been tricked into thinking that there was mutuality in the relationship between beekeepers and their bees? In this case, BVT reminds us that humans rank higher in importance than insects. Perhaps humans cannot resist exercising their power over insects simply because they can. But will that power always remain unchecked, even unquestioned? Or did the self-sacrificing nature of worker bees then extend to the beekeeper as a means of greater calling?

Noticeably, even the urban beekeeper's relationship to his or her bees is, at times, ambivalent. Some of the same informants, who wished to "work with the bees" to ensure their survival, ranked humans (themselves) over (their) bees when it came to bee-venom therapy. The notion of "sacrificing" an insect seems quite different from sacrificing a domesticated mammal, such as our informants' cats and dogs who often visited with us during our fieldwork. Where does the ethics of other species begin and end? To which species are we morally bound? This type of inquiry could be where the debates over sentience come in as a way to elevate animals to our level (of course, looking for sentience or intelligence in animals has now been criticized as entirely off point for its glaring humanness or human-centricness). We feel "bad" about killing bees, but we really don't think about what it feels like for them to be grabbed with our giant mammalian hands and pressed against our flesh until they fight to the death. There is a level of inhumane treatment of nonhuman animals that is somehow acceptable under certain circumstances but deemed cruel under other circumstances. Ending human suffering trumps not just the bees' suffering but also the bees' very existence. This is an argument we return to in our conclusions.

Sentience is the quality of awareness that a being possesses: a being that is sentient is endowed with feelings or capable of feeling sensation. These feelings include the capacity to suffer. Philosophers from Immanuel Kant to animal ethicists such as Tom Regan, argue over whether or not animals, including insects, are sentient.[31] We were not able to empirically observe what bees might have "felt" when we disrupted their homes, but they certainly reacted to these disruptions. Regardless of whether or not they were consciously agitated, or felt bothered, bees did at times buzz more loudly and at a higher pitch when we did hive inspections. Gauging the difference between bees reacting to environmental stimuli and feeling something about it is impossible for us. Not only are we unable to ask bees what they are feeling, but the idea of sentience itself is a multifaceted term that humans have difficulty defining for themselves. And humans do not agree about what sentience is. While we don't assert that bees have selfhood or are riddled with existential anxieties, their actions indicate that they are certainly aware of their surroundings and make decisions to adjust them if they are able. In terms of suffering, we suggest that certain practices—in particular, grabbing a bee and slamming it repeatedly against you until it stings and dies—are good starting points for exploring bee ethics, as predicated on sentience or intelligence as the case may be.

Military Uses of Bees

Just as the labor of bees has been harnessed for the production of food as well as medicinal treatments, historically, there is ample evidence that bees have also been used as weapons themselves. Beehives have been launched as bombs via catapults in several wars dating from before the 11th century, unsuspecting troops have been poisoned by strategically placed honey, and bees and beehives have been used as barricades to protect against enemy invasion. Bees were combat ready-engaged soldiers on the ground during battle.

Bees and human combat have been co-constitutive. In *Six-Legged Soldiers*, the entomologist Jeffrey Lockwood cites compelling evidence that suggests that bees are the "oldest tactic in biological warfare."[32] Indeed the word "bombard" comes from the Greek root *bombos*, meaning the buzzing of a bee. Perhaps, argues Lockwood, insects are considered good for warfare because, unlike other species, they are not considered to have the instinct of self-preservation. And certainly bees, with their hive mentality and work ethic, are perfect candidates for humans to layer meanings on the creatures as the ultimate self-sacrificers. Interestingly, in the 1930s and 1940s scientists who were members of the Nazi Party, such as the botanist Ernest Bergdolt, viewed the hive as embodying the sacrificial ideals of German national socialism. As the anthropologist Hugh Raffles observes, the "allegorical possibilities" of bees included "disciplined subjection to the well-being of the greater good, "self-sacrificial altruism," and "the dissolution of the individual in the anonymity of collective purpose."[33] The unique sociality of bees affords humans opportunities to both inscribe them with meanings and harness them for political purposes, such as creating weapons.

Propaganda and political metaphors aside, given that bees are part of the native environment (they are naturally "at the ready," so to speak), their memorable sting and psychologically menacing potential when en masse have been used as threats since the Paleolithic period. As soon as humans were able to throw beehives, there is evidence to suggest that they did so across cultures and historical epochs. Egyptian hieroglyphics provide some evidence of bees launched at enemies and invaders.[34] Technologies were also created and configured in concert with bees—just one of many instances of how we have creatively intervened with "nature" in the name of "culture." For example, the Tiv people of Nigeria designed a bee cannon, Mayans had a type of bee grenade, and Hungarians used bee boles. From the Romans to Scandinavians, bees have been used wherever they were found in local habitats. There are myriad examples of humans devising ways to catapult or lob hives at adversaries. Bees were more than props within ancient theaters of war. Indeed, they functioned as insect *actors* in the sense that they instilled enemies with fear and they performed a duty, albeit not of their own making.

Another way bees were used in wartime was through the attack on bee colonies in order to kill them off. Natural and political ecosystems have relied on the labor of bees and the distinctive portable bioarchitecture of their existence. The theory goes that if you destroy the bee population, then you disable the nation's agricultural output for homeland use and export revenue. One possible way to strangle bee colonies is through the introduction of harmful invader species such as the *Varroa* mite, which spreads a virus within the hive and is considered to be one of the several reasons for the emergence of CCD. *Varroa* mites are lethal parasites that thrive at the expense of honeybees. Lockwood argues that during the 1960s Cuba experienced an outbreak of *Varroa* mites in its bee population. Accusations were made that the United States was responsible for introducing the mites to Cuba. While these allegations are not supported by physical evidence per se, clearly killing hives and thus hampering agricultural production could be used as a strategy to weaken a nation's well-being in a tangible way.

Humans' reliance on bees' labor therefore creates another vulnerability to a nation: its food security. Arguably, a healthy and enduring relationship with bees is necessary. Dogs, horses, and dolphins can be weaponized or "commissioned" as agents of war, but bees are unique in that they can also be preyed upon, and their vulnerability places us at jeopardy. Infecting dogs, destroying horses, or killing dolphins would certainly affect a nation's cultural and ecological well-being, but this would not endanger a government's ability to feed and sustain its population. Notwithstanding, there are other notable and distinctive ways in which our relationship toward (and with) bees differs from other species during wartime.

Bees as both weapons in hand-to-hand combat and as a weakness to exploit through extermination are perhaps the most common ways to consider their role in entomological warfare. Bees have also been tested by entomologists for their possible use as a type of radar. In a sense, bees may be the insect version of the bomb-sniffing dog. Since the mid-1950s the U.S. Army "recruited bees" for hazardous waste duty, whereby they return to base after reconnaissance flying missions, picking up toxic agents on their fur. Bees fly in a radius of approximately five miles and forage in trees, flowers, and bushes, and it is not only pollen that they pick up on their fur and legs. As excellent navigators, bees return to their hives after excursions, and humans have taken advantage of these bee behaviors as well as the bees' olfactory system for detecting odorants. Bees have been "trained" to react to certain chemical agents used in bombs by fitting them with electronic equipment and observing their hovering patterns over particular chemical odors.[35] As nations attempt to identify the boundaries of the stockpiles of chemical weapons, there is no end to the use of bees as potential detection agents.[36] Bees might be useful at border crossings or even when determining if a nation is being dishonest in its use of certain chemical substances.

Most recently and over the past decade, the United States Defense Advanced Research Project Agency (DARPA),[37] with the Lockheed Martin Company and Sandia National Laboratories, has funded research to evaluate the use of honeybees as potential chemical and biological agent detectors. Hidden threats (chemicals you can't smell, see, or otherwise observe through human nonenhanced methods) require more unobtrusive means of detection—bees have been described as a perfect sentinel species. Called "nature's rugged robots" by Robert Wingo, a Los Alamos scientist, honeybees can be trained and deployed to provide data regarding the presence of toxins such as C4 or TNT.[38] Because of the ways bees sample the air, soil, water, and vegetation of a region through their bodies and practices, they are able to collect data in all the possible chemical forms, gases, liquids, and particulates. Bees are well equipped to be "border security sentries" and "combatants" in the war against agricultural bioterrorism.[39] The entomologist Jerry Bromenshenk's research at the University of Montana demonstrates the bee's ability to identify things by its chemical signatures with application to the identification of land mines.[40] Bees can also indicate whether the environment is "sick" from CCD and by actual retrieval of data to be mined—at little risk to the waiting human scientist. Just as humans are laboring to "save the bees," bees are unwittingly used to save the humans from the devastating environmental degradation of wartime activities. There is also research to suggest the creation of a cyborg bee, which may offer new potential in extra-post-bee capabilities with certain wartime/surveillance applications.[41] Honeybees can be militarized for covert ops, chemical surveillance, and classified missions—bringing home surveillance data that elude the collection abilities of any human.

Bee Welfare

Bees occupy a unique position as multifaceted creatures with flexibility for agricultural, medicinal, corporate, artistic, and wartime purposes. It is obvious how honeybees and their labor support local urban political economies, heal human bodies, and protect nation-states.

But what of the bees themselves? How can a creature, with so much potential for human benefit, be considered? Clearly, *Buzz* has examined myriad ways that bees and humans are enmeshed. In other words, in talking about bees and caring for them, humans deem what is important and meaningful to themselves and the bees. Contrary to a humanist interpretation, we posit that bees are not passive recipients of human direction; rather, bees react and respond to social environments with purpose and agency. They are not predictable, docile, and obedient subjects; they actively and creatively maintain their own network of action and interaction. They swarm, sting, navigate, collaborate, and decide. While bees communicate with what some scientists refer to as symbolic "language" through what is called "waggling"[42] (which indicates the distance and direction for potential food sources), they do not "speak" in ways that are audible to humans. Dogs bark and wag their tails, and cats may purr when they feel content and safe. Given their sizes, bees are much more elusive, and for the average person, they are particularly difficult to decipher. And even though we need them for agricultural production, we don't acknowledge their hard work. Unlike other species, bees are rarely "rewarded" for being "man's best friend" or incentivized by cozier accommodations or dietary treats. Their lives are not improved in any way by their participation in many human-motivated or mandated practices, particularly in the case of being used as a weapon. Is it up to people to speak on behalf of bees? The 1973 Nobel Prize winner Karl Von Frisch discovered "the language of bees" by decoding the waggle dance to indicate that it relayed information to other bees. As Raffles notes, "Von Frisch spoke for honeybees. And he made them speak. He didn't just give them language; he translated it."[43] Bees do not talk, and they do not have human feelings, but they do communicate symbolically. Given their utility, both within and outside of the context of war, the deployment of bees raises questions surrounding subjecthood and the ethical use of animals, including insects. The idea of bees as "subjects" is an aspect of the larger account of bees and war that is often absent from historical and biological analyses, as bees are implicitly exempt due to their status as mere insects. In this particular case, the notion of considering insects as an extension of the animal kingdom is worthy of debate.

The United States Animal Welfare Act of 1966 regulates the treatment of animals through federal law. Animals are defined in the following way:

> (g) The term "animal" means any live or dead dog, cat, monkey (nonhuman primate mammal), guinea pig, hamster, rabbit, or such other warm-blooded animal, as the Secretary may determine is being used, or is intended for use, for research, testing, experimentation, or exhibition purposes, or as a pet; but such term excludes (1) birds, rats of the genus Rattus, and mice of the genus Mus, bred for use in research, (2) horses not used for research purposes, and (3) other farm animals, such as, but

> not limited to livestock or poultry, used or intended for use as food or fiber, or livestock or poultry used or intended for use for improving animal nutrition, breeding, management, or production efficiency, or for improving the quality of food or fiber. With respect to a dog, the term means all dogs including those used for hunting, security, or breeding purposes.[44]

Insects do not rank and are never mentioned within the definition and therefore they are not considered as worthy of our consideration and protection. What are the ethical implications of using bees for human, any human, purposes? And as the philosopher Colin McGinn has argued, animals are only defined by their use value to human beings. In his essay about animal mortality, McGinn argues that animals must be granted selfhood because they have experiences. The metaphysical fact that unifies all species is that they are "all experiencing subjects."[45] Posthuman studies pushes us to consider the moral community of all experiencing subjects, and not just the warm and furry ones who lend themselves to illustrations in children's books.[46]

In the myriad ways that bees are used in commerce, medicine, and war, people seem to be grappling with their control over the "natural world" and human mortality. Many men, and some women, continue to attempt to control the "natural world" by using other species as beasts of burden, treatments, capital, weapons, weakness, spies, or theory in order to dominate the "wilderness" as well as discipline an unruly body, marketplace, competitor, environment, or situation. However, as with all species and things, it is not a unidirectional relationship. To some extent, it can be argued that bees control some aspects of human life. As discussed, our diets rely on their labor as pollinators, and a single bee has the capacity to inflict terror on an unsuspecting child or overly cautious or allergic adult. But bee "agency" as it were (i.e., the bees' free will and ability to decide on a course of action), is not entirely socially defined or relative to human context. For example, bees express preferences, travel with purpose, communicate directions, sting predators, and die en masse, not because we demand it of them, but because they are a purposeful and relatively autonomous species (as compared to animals we have traditionally domesticated). In many ways, bees, or the human interpretation and uses of them, are at the center of nature/

culture debates within the social and ideological space of war. They are a rudimentary and naturally occurring weapon as well as a complex cultural collective offering innovative thinking to the architects of war. The idea of bees as "bombs" communicates more about culture than it does about nature.

Furthermore, as a nonhuman screen, meanings from multiple perspectives are projected upon the bee—entomologists study bees for human purposes within individual research projects and sometimes within the context of larger political projects. These meanings of bees are relative in a way that runs the gamut from "healers" to "laborers" to "killers" to "friends." Their behaviors are individualized and also placed within communities that have a particular sociality. Depending on which human is doing the constructing, bees' sociality is interpreted to serve some higher purpose. So in addition to pollinating our almonds, apples, and dandelions, human ingenuity is inspired by bees' everyday lives.

Why should we care about bees? This is a difficult question that has economic, philosophical, and sociological implications. From a utilitarian standpoint, the honeybee's survival is intertwined with our own. Pragmatically speaking, it is in our best interest to keep the bee population healthy for its agricultural labor and in keeping with environmental sustainability. Simply put, and as much as we are skeptical of some of the co-optation, being green is good (for the economy and also from an ethical standpoint, and for bees in the reduction of pesticides and the creation of new habitats). But we can also learn about ourselves through our interactions with bees. How we elevate (theoretically) and simultaneously diminish their utility exposes inconsistencies in our relationships with animals and insects and to nature, broadly speaking. Bees can be positioned at the center of posthuman debates, where we move beyond the human as the center of all thought and action. Can bees be a companion species? As Donna Haraway suggests, if we examine our own genealogy and deeply consider how we are indeed a "tangled species" coming to being while inextricably being linked with other species, we may figure ways out of the Enlightenment thinking that has limited our analyses and practices. Instead, insisting how we are superior to bees and therefore free to exploit them, we ask how humans have emerged in collaboration with bees—our food, our agriculture, our health, our wars? Companion species perspectives enable us to ask questions about the interactive relationships of human species to honeybee species and about the messiness between us.

Human-bee collaborations spawn capital enterprises. As the beekeepers express, it is exciting to use raw materials, commune with the bees, harvest, create, and sell. The thrill of creation and commerce adds to the intercorporeal buzz generated by close encounters with bees. Taking our lead from Haraway, perhaps our relationships with honeybees can be guided by the ethics of mutual use and mutual exploitation. At Honey Fest, urban beekeepers raise awareness about bees' health and habitats, demonstrate careful standards of animal/insect husbandry, and show how they deeply care about their bees. It is through the health and sustainability of their beehives that they can engage in the ecological mesh of the city. When executed mindfully and with attention to the longevity of both species, mutual regard has been beneficial to both.

We want to be careful not to simply argue that bees should be given a "voice" because of their use value to us. Rather, while that might indeed be true in the case of the academic production of knowledge (i.e., articles and monographs for professional advancement), we are also saying that perhaps we should think about ways bees may be especially communicative. Bees may be the postmodern "canary in a coalmine," harbingers of environmental collapse. Bees are also interdependent species. We want to explore the ways that the bee, as an insect or othered animal species, can help us interrogate the perceived boundaries between animal and human. Because bees are so entangled with our everyday lives and bodies—from dietary consumption to the use of pollen and honey as medical curatives—they physically rupture the line between us (humans) and them (animals). They make us feel good emotionally (through communing with them or buying Burt's products), and in some cases they keep us alive and defend us from harm. Even when deployed as military agents or literary metaphors, bees are not simply other objects. The complex and contradictory ways in which we harness and live with bees illustrate that the boundaries between animals and humans are porous, not absolute. In this way, the bee can challenge us, in theory and practice, to reconsider the "natural" hierarchical order implied within notions of the animal kingdom and humankind.

8

Becoming Bee Centered

Beyond Buzz

We entered the field with the desire to understand our fellow urbanites in their yearning to "connect with nature" through gardening, chickens, bees, window boxes, community-supported agriculture, and metropolitan farmers' markets. In sociology speak, beekeeping was a "doable project." We established our entrée with a diverse community of beekeepers eager to share their stories. We had access to a variety of field sites to get at the range of variation of the social phenomenon of urban beekeeping. It is a timely subject and socially relevant. It didn't require much financial investment—always a consideration for public college professors at financially strapped institutions.

Leaving the "field" is usually an emotional experience when saying goodbye to informants, hoping to do their stories justice, and wishing the planned analysis is fascinating and cogent, readable, and relatable. Also there is the disciplinary imperative not to "go native" or "drink the Kool-Aid"—you don't want to lose the distance from the field so that critical, healthy skepticism keeps you on your intellectual toes.

Good fieldwork does create social relationships with informants, and sometimes these relationships can be tricky. The interpersonal expectations and emotional engagements are sometimes challenging to navigate. Certainly in the past, our informants have affected us; something about their human fragility and vulnerabilities, reflecting our own, makes their stories deeply gripping to us. Similarly, from the scientist to the backwards beekeeper, we feel attached to this community. Their practices offer yet another illustration of human creativity and a quest to make meaning out of life. Urban beekeepers are a very interesting tribe who deliberately and intentionally pursue interspecies relationships with insects—their relationships emerge from soul-searching, altruism, commerce, activism, and community building. For us, ongoing human relationships with these informants have been varied and interesting—we see informants intermittently at farmers' markets, bars, and "green" events in the city. We receive email updates about social events and honey-related celebrations. All our human informants have affected our lives.

What we didn't account for was the way the bees would change us. There is something invaluable to be gained from the buzz of intermingling with bees. As Donna Haraway reminds us, "Trans-species encounter value is about relationships among a motley array of living beings."[1] The bees persist to buzz in many unexpected ways; they distract, engage, and captivate us. They also linger. As we have delved deeper into extended social observation of these urban insect dwellers, they have come into our homes, our dreams, and our everyday musings. We must consider their everyday lives in the borough, the city, the metropolitan area, the state, the nation, the continent, and the planet. We have talked at length about how we feel angry on behalf of the bees for the callous and exploitative practices that prey upon their generosity. And it has even forced us to think whether some urban beekeepers are unwittingly keeping bees to ensure their own indispensability to the bees. Bees might or might not actually need human assistance, but they provide humans with such purpose and self-importance.

Perhaps the largest "finding" from our research is that we are indeed deeply intimate and interdependent with the bee as a species and the bee as a fellow urban dweller. However, we would not have realized the interconnectivity of bees had we not spent three years with them (see figure 8.1). Now we see bees everywhere among us. When we are in the grocery store comparing different types of honey to buy, or when we notice signs of spring on tree limbs, we think about their lives and work. As seasons change to cooler temperatures, we have even texted each other pictures of bees with the caption "even in the frosty fall, this girl is searching over the cement." It's not simply a jar of honey; it's the bees' honey. It's not a magnolia blossom but a bee pollination source. Bees are no longer an absent presence. We wonder how we took this species for granted and what that says about us as individuals and as a culture. Understanding our relationship with bees on a deeper level has led to many discussions about other species—we have talked about becoming vegan, the personalities of pet dogs, whether elephants make art, and the ethics of killing spiders. Our humanness has been exposed

Figure 8.1. Worker bees making honey. (Photo credit: Mary Kosut)

to us, and now we are turned on our heads. We more reflexively observe the nonhuman world from a new "less-human" angle and it's because of the bees. How did they slip outside of our consciousness for so long?

Ultimately, we are leaving the field changed not only by our human subjects but also by the bees. And while we do not necessarily have a rigid new code of ethics to follow in our own everyday lives, we acknowledge new layers of complexity that such a consciousness-raising experience creates. Bees will stay with us from this project as a species to consider in our everyday lives and as a reminder that we are not as self-sufficient as we might believe.

Capturing Bees

Bees are insects that do not stay within neat categorical boundaries of human syntax—no matter how humans try to control or manipulate them. Rigid social taxonomies cannot contain the magnificence of the species. They are insects, but not typical bugs or pests who are of limited obvious value to human life. Their yellow and black, double-winged, shiny, and fuzzy abdomens zip through bucolic pastures and urban decay. They are "disappearing" and yet visually everywhere, from media reports about their demise to school calendars or special "guests" at classroom visits. Bees are autonomous but part of animal husbandry since practices of domestication attempt to tame their activities in the pursuit of human propensity for predictability. They live freely yet they are enslaved to labor on industrial scales and for rooftop hobbyists. They reproduce across racial boundaries although their migration practices are legislated by governments. We have established intimacies and relationships with bees and hives and yet we have never pet them or gazed into their eyes. They do not sleep in our beds, or sit on our laps, and still our bodies have swelled from their injection.

And even though this personal experience with an individual bee might be lacking or unnoticed, bees figure prominently in human imaginations. Through the stories we tell and the phrases we use to express ourselves, we layer meanings over bees just as we interpret bees' lives as evidence of natural logics. Their activities are narrated by humans to explain or justify the order of things—naturally—be it reproduction (the birds and the bees), industry (busy as a bee), gendered performance (the queen bee, the bee's knees), or obsessive compulsions (bee in one's bonnet). Like other humans before us, we use and understand bees as signifiers; they are "traded" in a symbolic way. They amount to different types of currency, including the monetary value of honey and other bee substances, but the bee is also a visual symbol that is easily read. Bees are trafficked as codes—they are innocent, fertile, busy, industrious, self-sacrificing, communal, directed, threatened.

Bees are wholly othered in the animal kingdom, in the realm of the insect/pest, and in our everyday lives. They are creatures in a betwixt and between status—not really animal enough due to their minuscule size and relative autonomy in human relationships, but at the same time so close to us. They push boundaries of sociality and breach flesh, buzzing in our ears and forcing little dances of human jerkiness. We ingest the bee, digest its produce, invest our emotions, and extract money and labor from it in an unequal interspecies exchange. Within the framework of animal/human social theory and interspecies research, where does an insect fit? As we have argued throughout this book, bees are flexible in their role as the model insect, ceaselessly working for the benefit of the larger social group, or the cyborg mascot, modified by human technology for improved performance, output, and utility.

Because of this "othered," foreign, alien location, bees are good subjects for fieldworkers and theorists precisely because they are so confounding. Insects are recognized as members of the animal "kingdom," but in fact they dominate it as a species in terms of numbers and kinds. But from a human perspective, insects are often seen as living in another space and place—not an animal kingdom but an insectlandia, or parallel universe that borders our own. Insects fly or crawl or lurch out of insectlandia, an unknown world, into our human world. And our reaction usually is to shoe, swat, or squish them, because we can't deal with *them*—they are pests. Theoretical nature/culture boundaries between "us and them" often obscure the complex exchanges and interdependencies between species. The us/them boundary is made more porous by our eating their honey (or the vomit they have expelled from their stomachs) or by having their stinging venom circulate in our bodies. With the emergence of interspecies ethnographies, insects, other invertebrates, and invisible organisms are now recognized as legitimate research subjects, rather than as creatures on the margins of insectlandia. We see that these others are indeed among us and always have been.

Their properties are contradictory, and maybe because the bee is such an anomaly, it offers another chance to interrogate how we conceptualize animal-human relationships and our role as ethnographers covering nonhuman subjects. Bees are hard to pin down because they are not static; they are always working and on the move. And whenever bees seem to be slightly out of control, humans claim to be able to tame them. Perhaps bees are a semiferal species, and that is what attracts urban dwellers because it allows them to bring a little wildness into the city. Humans purposefully invite bees to their rooftops, backyards, and urban plots because of a growing interest in green lifestyles. Just like living in a high-density city, there is the ubiquitous air of danger in keeping bees, adding to the buzz. Maybe urban beekeepers have an advantage in performing these tasks of hive checks, sugaring bees, and harvesting honey in that New Yorkers must be a little tough, have moxie, and be on guard in their urban dwellings.

Bees are certainly feared by humans, and this plays into their affective associations. They are also loved and respected. Part of the vibrational force, the hum, and the contact high that one gets from bees is ecologically productive. Buzzing creates an ambience that touches humans on a sonic level. The theorist and dubstep performer Steve Goodman has explored the ontologies of vibrational force. As Goodman writes, "Affective tonalities such as fear, especially when ingrained and designed into architectures of security, can become the basis for a generalized ecology, influencing everything from microgestures to economics. As such, and unlike an emotional state, affective tonality possesses, abducts, or envelops a subject rather than being possessed by one."[2] In other words, Goodman explores how sound can get on your nerves and cause you to act in certain ways. The buzz can drive you crazy, make you move your body, and interrupt your ability to think. Yet the buzz encompasses our other sensory capacities—bees stimulate our taste buds, trigger our sense of smell, and keep us feeling their touch (sting) for days. When a bee enters our line of vision, the affective buzz awakens a complex palette of emotional tones and timbres.

In addition to being drawn to the buzz, some urban beekeepers see it as their calling to take on ecopolitical activism as it benefits local ecologies and potentially reverses environmental crises. These people do experience deep gratitude and kudos from their fellow humans for taking on this ostensibly onerous and dangerous job. Indeed, beekeeping is a virtuous hobby, more so than knitting or beer brewing, where humans work to save a vulnerable species and in the process perhaps help to save the planet. Beekeepers universally spoke to us about how their relationship with bees is an integral part of their identities—and it does not go unnoticed by the beekeepers that their hobby has cultural cachet. It is a hip urban trend that, similar to skateboarding or graffiti writing, moves individuals beyond their own comfort zones. In fact, our informants, by and large, were deeply dedicated to their bees, and their devotion goes beyond any possible notoriety or cocktail party banter that the hobby might provoke. Collectively we pin our hopes on beekeepers as a sign that human and nonhuman lives will prevail in environments choked by human progress, consumption, and desire. Perhaps humans have not simply domesticated bees; instead, they have created a new place and social space for the city farmer.

Throughout this work, we shared our struggles in attempting to understand the ontology of beeness and we queried whether or not beeness can exist in a world that has been stepped on by humans. Responsibly attending to our human informants, we also wished to represent the bees. We grappled with how to best accomplish that difficult task. Through photographs? Through the English language? Through our informants' descriptions of their labor? And what of decentering the human? It is very cutting-edge to say that we are privileging the world of the bee, when we are not quite sure what that means precisely in actual enacted social scientific research. Audio recordings of buzzing bees might sound nice in our classrooms but we still have to narrate what we imagine the bees are saying or doing in their buzz and why we should pay attention. We could bring bees to public lectures to show what they look like in a comb, but we would have to explain what they were doing and why.

Buzz argues throughout that human-bee interactions are in no way reciprocal. Humans take from bees, trample on their habitats, and kill them for "safety's sake." However we run into challenges when attempting to theorize a more reciprocally fair or just relationship between human and insect (as opposed to human and animal). Humans have an instrumental relationship to bees. Bees are unwittingly and forcefully employed by the government, corporations, medical industries, and holistic healers. Bees are thrown (sometimes literally) into systems of commerce, political and metaphoric battles, and semiotic systems. When humans make contact with bees it unearths or illuminates how destructive and dominating we are. Is it possible that urban beekeepers are trying to remedy the atrocities of previous violent acts toward bees? Some of us may be making up for the damages inflicted by our species. This atonement is a form of penance for anthrocentric behaviors, but it may always be framed in liberal humanistic terms of freedom, oppression, or liberty.

Despite many humans trying to convince us otherwise, we have come to believe that our species needs the bees but they don't need us. Our food chain depends on bees' labor to sustain human life—and the reverse is not true. We are deeply intertwined with bees in a multispecies relationality that cannot be disentangled. So what might it look like to have an ethical practice toward bees? Based on our own experiences with and analysis of the bee we offer two possible options for our future relationship with bees: *radical disengagement* and *ethical engagement*.[3]

With the first option, radical disengagement, we could establish a more veganlike relationship with the bees. Vegans practice a lifestyle that endeavors to abolish animal exploitation and vegans do not use any animal products for food (including eggs, honey, dairy), clothing, pharmaceuticals, or cosmetics. Radical disengagement would mean we don't take anything from bees and we let them roam, live, and pollinate what and where they want. There would be no honey harvest; there would be no bee-sting therapies. Bees would swarm wherever they chose and produce as much honey as they'd like. We wouldn't breed them or limit their migration patterns or facilitate new paths of migration. We wouldn't overmedicate bees with antibiotics for mites, parasites, or other pests. We would not use them for nation building or military reconnaissance. All of these purposeful changes are aimed directly at undoing human interventions into bees' everyday lives. Furthermore, we would stop genetically modifying crops and using pesticides and insecticides. On a larger scale we could stop our destruction of ecological habitats, end industrial (air, land, water) pollution, and curb industrialization. We could radically change our everyday human lifestyles, including riding bikes and using public transportation, using solar and wind energy, limiting consumption, supporting organic agriculture, adopting vegetarianism, and eating locally.

There are drastic consequences for such a radical change. Certain crops might fail without forced pollination by bees, leading to changes in human dietary options and nutrition. There would be no honey for drugs, cosmetics, sweeteners, medication, salves, and industrial food production. Left to their own breeding practices, new hybrid bees would most likely emerge and bees would take on new ethnic makeups. So-called Africanized bees might migrate northward more quickly and humans may need to develop strategies to not get into their way. While there would be no need for beekeepers, as we currently understand them, we would still need to cultivate human experts who can guide us in the newly formed human-bee relationships. Radical disengagement will certainly not be better for most humans, but possibly it would be for the bees. Many species would benefit from large-scale structural changes that limited pollution and habitat destruction. Of course, there are some critters who are adept at living with humans and our various forms of pollution; bedbugs, rats, dinoflagallates, chytrid fungi, and Xenopus frogs, to name a few. Species such as these would likely decline if human impacts on the landscape were minimized. However, as pragmatists, we think this scenario is highly unlikely. Too many people profit from bees and bee products. Such radical ecological changes are also so challenging to mount that it leads to a sense of helplessness and futility, generally resulting in no policy or behavior changes.

The second option, ethical engagement, involves creating more deliberate and thoughtful working conditions for bees that consider their health and well-being. In New York City, our informants explained how the life span of a worker bee is between one to four months, a queen is about two years, and a drone is about forty days. This relatively short time in human terms may persuade some that there is no need to consider the working conditions of bees. However, entomologists, beekeepers, and invested human others could join together to establish the best practices for the species. Furthermore, when considering honey harvesting, the bee must be part of this equation that does not simply ask what is the bare minimum of the honey necessary for the hive to survive, but rather what is in the bees' best interest? Honey would still be harvested and available but the standards for extraction would be strictly controlled, possibly with less for human consumption and use in medical practices. A percentage of the proceeds from the sale of honey could be reinvested into creating native pollinator gardens and building hive boxes for bee colonies.

When it comes to treating bees for parasites, most specifically *Varroa*, minimal and cautious treatment might be in the best interest of a colony that is suffering from losses. In the practices of ethical engagement, pollination on an industrial scale could still happen but would likely be more expensive to mount, leading to higher prices for food. In the case of commercial beekeeping, bees' labor would not be viewed as exhaustible and they would be given a range of plants, rather than one crop only, to pollinate. Standards for transporting bees to industrial farms as well as best practices for workdays might also establish a baseline for bees' optimal labor for their lives and health. For example, our informants with experience in commercial pollination argue that bees require diversity in agriculture, and monoculture is not conducive to bees' health. Enabling bees to have breaks in pollinating apples or blueberries for weeks—perhaps with a weeklong break for bee-directed pollination—is worthy of investigation. Furthermore, rather than blindly breeding certain types of bees that are perceived as desirable to humans, we are more cautious and aware of what these genetic practices may do to bees themselves. In a nutshell, we seek to become more bee centered.

Ethical engagement relies on modifications in our practices and relationships with bees; it seems to us that this speaks to contemporary notions of sustainability. We don't disconnect from them as in the radical disengagement scenario, but rather, we seek to *engage* with bees in a manner in which they are not taken for granted as an inexhaustible natural resource. We reposition bees in relation to us, not as mere insects but as a highly valuable species that must be protected, much like trees have come to be understood as a resource to be protected from deforestation. Like the depletion of other natural resources, we know that severe consequences will extend well beyond CCD if we do not act ethically toward the bees. We acknowledge that bees are a vital source of life across the planet, for humans, plants, and other animals. And in the interest of real sustainability, we must work to modify our lifestyles in myriad forms for the sake of bees and humans alike. We need to become aware of our ecological inertia and complacency and be more critical of buying into symbolic green consumerism.

As much as we are drawn to both radical disengagement and ethical engagement, we must offer one final caveat. We do not want to suggest a harmonious and Eden-like return to a state that would be potentially bucolic and healthy for bees alone. As Donna Haraway reminds us, we as humans have already irrevocably transformed the world. We have had more negative impacts on bees than they have had on us. Hierarchically positioned at the top of the animal kingdom, we have fashioned the world to suit our desires, resulting in the direct exploitation of natural and animal resources. At the same time, humans naively or perhaps arrogantly believe that when we become enlightened about the plight of a particular species, we will be able to make conditions better for them and can remedy our human mistakes. This fixing may take the form of letting bees go "wild" and stopping the use of pesticides as proposed by backwards beekeepers such as Sam Comfort and Kirk Anderson. We hope that somehow the bees' lives will be improved if we "let bees be bees" and remove ourselves from the relationship. Importantly, on a much larger scale, humans have changed the planet through global warming; industrial pollution of air, water, and land; and many other means. To eradicate the use of pesticides, ban cars, make bee gardens, or let bees swarm naturally may not solve the ecological dilemma—such changes, no matter how altruistic or radical, may not be enough. After all, our ecological awareness and burgeoning interspecies egalitarianism have actually been born out of centuries of human innovation and domination of the biosphere.

We cannot simply intervene or unintervene and allow animals (domesticated in particular) to just be free because our consciousness is raised and our guilt is heightened. Certainly, we cannot open the doors to our apartments and free our dogs or unleash them on the street—the world humans have created would be quite uninhabitable for them. A comparable analogy is to imagine the consequences of freeing animals from a zoo. However, insects such as bees are different because they are semiferal and semidomesticated. Technically, we could free bees because they don't need us to "own" them in the same way that dogs or cats do. We have yet to make bees wholly dependent (or interdependent to us). What is so confounding and interesting about insects, especially social insects, like bees, is that they thrive well together, making them distinct from other domesticated species. We could fall away from the bees' lives quite easily, as we are mostly irrelevant to them for survival. Yet we cannot let ourselves completely disengage from bees because the consequences would tangibly affect us as a species. We do not want bees as companion species to cuddle with, but at this point in history we do need them to continue living in the manner we are accustomed to.

Finally, after our three years of fieldwork and two lifetimes of being deeply dependent on bees, we are still not able to definitively say what makes the bees happy and what their deepest preferences might be. We can surmise, however, what it is to die and suffer as a bee. We have seen them thrive in bucolic settings and in the midst of the flotsam and jetsam of New York City. Bees are obviously adaptable, but all living organisms have certain limits, breaking points, and vulnerabilities. We are inviting a nonhuman species to join us in urban spaces, and bees do make our cities better places to live. We must work harder to consider the bee and its well-being.

NOTES

CHAPTER 1

1. What measures are planned to help contain or control the bees? If plans do not include containment (e.g., within screening), what safety measures would be utilized?

There will be no permanent structure to control the bees. We are specifically not asking for funding to build beehives due to the lack of ability to tend to the hives by student/faculty members throughout the year. Rather we choose a plot as a means to attract bees and enable them to have a source of native pollination. However, we believe there are already bees in the area, though we have not conducted a census; in particular bees are drawn to dandelions, which are abundant on our campus. As far as safety measures, the native pollinator project is part of the Center for Biodiversity and Conservation and the Greenbelt Native Plant Center: see http://greatpollinator-project.org/aboutus.html. This suggests that planting native plants does not create any greater risk from bees—in fact there is some evidence to suggest that a native habitat would concentrate bees in an area as a food/pollen source. We will clearly label the plot as a bee pollinator project plot so that individuals are aware of the probable presence of bees.

—Estimate of the amount of bees expected?

Since we have not conducted a census it is difficult to get an accurate sense of the size of the bee population or the diversity of species of bees on the Purchase campus. Based on evidence from the Brooklyn Botanic Garden's pollinator project, it is possible that there could be anywhere between forty thousand and eighty thousand bees living near or around the garden plot. It is unclear if the native pollinator plot would increase this population. We are not aware of the bees' relative health at Purchase and if they are in any way suffering from Colony Collapse Disorder.

—What training is planned (initially and ongoing) for those who will be "beekeepers" or otherwise work with the bees?

The great pollinator project has all its materials and helpful videos and tracking sheets, and identification illustrations available online. Interested students and faculty will be asked to print and download instructions from the website. Individuals can set up independent accounts with the pollinator project and provide their data to the group directly. Furthermore, we will have an organizational meeting at the beginning of each semester to explain the garden to interested Purchase community members and show them how to go about becoming equipped to be citizen scientists. This project requires very limited organization, as it is more a do-it-yourself scientific experience and one that hopefully benefits the bees first and foremost as well as their interested humans.

2. Monica Casper and Lisa Jean Moore, *Missing Bodies: The Politics of Visibility* (New York: NYU Press, 2009).

3. See Tom L. Beauchamp and James F. Childress, *Principles of Biomedical Ethics*, 3rd ed. (New York: Oxford University Press, 1989).

4. Lucette Matalon Lagnodo and Shelia Cohn Dekel, *Children of the Flames: Dr. Josef Mengele and the Untold Story of the Twins of Auschwitz* (New York: Penguin, 1992).

5. The Belmont Report laid out three general ethical principles that should govern human subject research: beneficence—to maximize the benefits of science, humanity and research participants and to avoid or minimize risk or harm; respect—to protect the autonomy and privacy of rights of participants; and justice—to ensure the fair distribution among persons and groups of the costs and benefits of research.

CHAPTER 2

1. There are four thousand species of bees in North America.

2. Some native plants that are known to be good for bees include daisies, asters, lavender, mint, sunflowers, heather, sedum, sage, borage, echinacea, salvia, roses, snapdragons, verbena, buddleia, poppies, marigolds, fennel, and Queen Anne's lace.

3. Kaija Helmetag, "Think of Them as Your New Pets," *New York Magazine*, 2010, http://nymag.com/guides/everything/urban-honey/66172/ (June 9, 2011).

4. Rose-Lynn Fisher, *Bee* (New York: Princeton Architectural Press, 2010), 105.

5. Ibid., 106.

6. Joshua Brunstein, "Beekeepers Keep the Lid On," *New York Times*, 2009, http://www.nytimes.com/2009/06/21/nyregion/21ritual.html?ref=bees&_r=0 (accessed October 20, 2012).

7. Ibid.

8. Christopher Mele, *Selling the Lower East Side: Culture, Real Estate and Resistance in New York City* (Minneapolis: University of Minnesota Press, 2010), 13.

9. Jennifer Wolch, "Anima Urbis," *Progress in Human Geography* 26, no. 6 (2002): 722.

10. Susan Dominus, "The Mystery of the Red Bees of Red Hook," *New York Times*, 2010, http://www.nytimes.com/2010/11/30/nyregion/30bigcity.html (accessed July 11, 2011).

11. William Longgood, *The Queen Must Die and Other Affairs of Bees and Men* (New York: Norton, 1988), 230.

12. Thomas Seeley, *Honeybee Democracy* (Princeton, N.J.: Princeton University Press, 2010), 6.

13. Ibid., 4.

14. Freya Mathews, "Planet Beehive," *Australian Humanities Review* 50 (May 2011), http://www.australianhumanitiesreview.org/archive/Issue-May-2011/home.html (accessed October 22, 2011).

15. See Monica J. Casper and Lisa Jean Moore, *Missing Bodies: The Politics of Visibility* (New York: NYU Press, 2009), for an exploration into how missing human bodies require new ways of seeing.

16. Agricultural Marketing Resource Center website, http://www.agmrc.org/commodities__products/livestock/bees_profile.cfm (accessed August 14, 2011).

17. "The Global Silk Industry: Perception of European Operators toward Thai Natural and Organic Silk Fabric and Final Products," http://www.fibre2fashion.com/industry-article/38/3793/the-global-silk-industry1.asp (accessed December 8, 2011).

18. Agricultural Marketing Resource Center website, http://www.agmrc.org/commodities__products/livestock/bees_profile.cfm (accessed August 14, 2011).

19. Holly Bishop, *Robbing the Bees: A Biography of Honey, the Sweet Liquid Gold That Seduced the World* (New York: Free Press, 2005), 238.

20. Ibid., 239–240.

21. Donna Haraway, *When Species Meet* (Minneapolis: University of Minnesota Press, 2008).

22. Ibid., 235.

23. Jen Wrye, "Beyond Pets: Exploring Relational Perspectives of Petness," *Canadian Journal of Sociology* 34, no. 4 (2009): 1037.

24. Peter Singer, *Animal Liberation* (New York: HarperCollins, 1975).

25. The behaviors of certain animals, in particular, elephants, dolphins, octopi, and starfish, have been used as evidence to assert that these species exhibit intelligence. The logic goes that if they are intelligent then they are sentient and in effect closer to humans. Thus, they have rights as a species. Some argue that this perspective is problematic because intelligence is a human construct and very difficult to measure.

26. Joel S. Savishinsky, "Pet Ideas: The Domestication of Animals, Human Behavior, and Human Emotions," in *New Perspectives on Our Lives with Companion Animals,* ed. Alan M. Beck (Philadelphia: University of Pennsylvania Press, 1983), 112–131.

27. Judith Siegal, "Companion Animals: In Sickness and in Health," *Journal of Social Issues* 49, no. 1 (1993): 157–167.

28. Lisa Wood, Billie Giles-Corti, and Max Bulsara, "The Pet Connection: Pets as a Conduit for Social Capital?" *Social Science & Medicine* 61, no. 6 (2005): 1159–1173.

29. Clinton R. Sanders, "The Animal 'Other': Self-definition, Social Identity and Companion Animals," in *Advances in Consumer Research*, Vol. 17, ed. Marvin E. Goldberg, Gerald Gorn, and Richard W. Pollay (Provo, Utah: Association for Consumer Research, 1990), 662–668.

30. Arnold Arluke and Clinton Sanders, foreword to *Between the Species: Readings in Human Animal Relations* (Boston: Pearson, 2008), xix.

31. Ibid., xviii.

32. Steven Best, "The Rise of Critical Animal Studies: Putting Theory into Action for Animal Liberation into Higher Education," *Journal for Critical Animal Studies* 7, no. 1 (2009): 9–54.

33. Helena Pedersen, "Release the Moths: Critical Animal Studies and the Posthuman Impulse," *Culture, Theory and Critique* 52, no. 1 (2011): 65–81 (quote on 72).

34. Ibid., 67.

35. Eduardo Kohn, "How Dogs Dream: Amazonian Natures and the Politics of Transpecies Engagement," *American Ethnologist* 34, no. 1 (2007): 5.

36. S. Eben Kirksey, and Stefan Helmreich, "The Emergence of Multispecies Ethnography," *Cultural Anthropology* 25, no. 4 (2010): 545–576.

37. Ibid., 545.

38. Edward O. Wilson, *The Ants* (Cambridge: Harvard University Press, 1990). This work is a collaboration with Bert Holldobler.

39. See, for example, Marshall David Sahlins, *The Use and Abuse of Biology: An Anthropological Critique of Sociobiology* (Ann Arbor: University of Michigan Press, 1976).

40. For the anthropologist Bruno Latour, there's a "democracy of objects" in which no body or no being or no thing can possess ontological priority over anything else. However, for Latour, an object can be "strong" or "weak," depending on the number of relations it maintains at any given moment.

41. Timothy Morton, *The Ecological Thought.* (Cambridge: Harvard University Press, 2010).

42. Karen Barad, *Meeting the Universe Halfway: Quantum Physics and the Entanglement of Matter and Meaning* (Durham: Duke University Press, 2007), 89.

43. As the media scholar Douglas Kellner argues, "Media culture helps shape both an individual's and a society's view of the world and deepest values . . . as well as who and what are seen as threats and enemies, creating, in some cases, sharp divisions between 'us' and 'them.' " From "Cultural Studies, Multiculturalism, and Media Culture," in *Gender, Race and Class in Media,* ed. Gail Dines and Jean M. Humez (Thousand Oaks, Calif.: Sage, 2003), 9.

44. Kirksey and Helmreich, "Emergence of Multispecies Ethnography," 545.

45. Donna Haraway, "Encounters with Companion Species: Entangling Dogs, Baboons, Philosophers and Biologists," *Configurations* 14 (2006): 97–114.

46. Hal Herzog, *Some We Love, Some We Hate, Some We Eat: Why It's So Hard to Think Straight about Animals* (New York: Harper, 2010), 279.

CHAPTER 3

1. U.S. Environmental Protection Agency website, http://www.epa.gov/opp00001/about/intheworks/honeybee.htm (accessed August 14, 2011).

2. Kirk Johnson, "Scientists and Soldiers Solve a Bee Mystery," *New York Times,* October 2010,

http://www.nytimes.com/2010/10/07/science/07bees.html?_r=2&src=ISMR_HP_LO_MST_FB (accessed August 15, 2011).

3. Donna Haraway, *When Species Meet* (Minneapolis: University of Minnesota Press, 2007), 46.

4. http://globalclimatechange.wordpress.com/2007/04/20/einstein-on-bees/ (accessed July 7, 2010).

5. http://www.beeguardian.org/ (accessed July 8, 2010).

6. Elizabeth Kolbert, "Stung: Where Have all the Bees Gone?" *The New Yorker*, 2007, http://www.newyorker.com/reporting/2007/08/06/070806fa_fact_kolbert?currentPage=4 (accessed September 11, 2012).

7. Jessica Hamzelou, "Where Have All the Bees Gone?" *The Lancet* 370, no. 9588 (August 25–August 31, 2007): 639.

8. Emily Martin, *Flexible Bodies: Tracking Immunity in American Culture from the Days of Polio to the Age of AIDS* (Boston: Beacon Press, 1994).

9. Kirk Johnson, "Scientists and Soldiers Solve a Bee Mystery," *New York Times*, October 2010, http://www.nytimes.com/2010/10/07/science/07bees.html?_r=1&scp=1&sq=bees,%20fungas%20tag%20time%20virus&st=cse (accessed August 15, 2011).

10. Ibid.

11. Ibid.

12. Sainath Suryanarayanan and Daniel Lee Kleinman, "Disappearing Bees and Reluctant Regulators," *Issues in Science and Technology Online*, http://www.issues.org/27.4/p_suryanarayanan.html (accessed October 3, 2012).

13. Ibid.

14. Kim Flottum, *The Backyard Beekeeper* (Beverley, Mass.: Quarry Books, 2010), 191.

15. Steve Kroll-Smith and H. Hugh Floyd, *Bodies in Protest: Environmental Illness and the Struggle over Medical Knowledge* (New York: NYU Press, 1997).

16. Katherine Harmon, "Zombie Flies May Be Killing Honey Bees," *Scientific American*, http://www.scientificamerican.com/article.cfm?id=body-snatching-flies (accessed May 26, 2012).

17. Gosia Wozniacka, "Study: Parasitic Fly Could Explain Bee Die-Off," January 2012, http://cnsnews.com/news/article/study-parasitic-fly-could-explain-bee-die (accessed January 4, 2012).

18. Harmon, "Zombie Flies May Be Killing Honey Bees."

19. Kyle Bishop, "Dead Man Still Walking: Explaining the Zombie Renaissance," *Journal of Popular Film and Television* 37 (2009): 16–25.

20. Wozniacka, "Study: Parasitic Fly Could Explain Bee Die-Off."

21. Katherine Eban, "What a Scientist Didn't Tell the *New York Times* about His Study on Bee Deaths," published October 2, 2010, in CNNMoney.com, http://money.cnn.com/2010/10/08/news/honey_bees_ny_times.fortune/index.htm (accessed October 22, 2010).

22. Ibid.

23. Ibid.

24. Charles Martin Simon, "Principles of Beekeeping Backwards," originally published in *Bee Culture*, July 2001, http://www.beesource.com/point-of-view/

charles-martin-simon/principles-of-beekeeping-backwards/ (accessed September 19, 2011).

25. Ibid.

26. Jan Suszkiw, "New Antibiotic Approved for Treating Bacterial Honey Bee Disease," 2005, http://www.ars.usda.gov/is/pr/2005/051219.htm (accessed September 19, 2011).

27. Michael F. Potter, "Parasitic Mites of Humans," http://www.ca.uky.edu/entomology/entfacts/ef637.asp (accessed September 30, 2012).

28. Sarah Franklin and Margaret Lock, *Remaking Life and Death: Toward an Anthropology of the Biosciences* (Santa Fe, N.M.: School of American Research Press, 2003), 102.

29. Matthew Immergut, "Manscaping: The Tangle of Nature, Culture, and Male Body Hair," in *The Body Reader*, ed. Lisa Jean Moore and Mary Kosut (New York: NYU Press, 2010), 287–304.

30. George Sessions, *Deep Ecology for the 21st Century* (Boston: Shambhala Press, 1995), xxi.

31. Hugh Raffles, "Sweet Honey on the Block," *New York Times,* 2010, http://www.nytimes.com/2010/07/07/opinion/07Raffles.html?ref=urbanagriculture (accessed February 2, 2011).

32. Kristina Shevory, "The Beekeeper Next Door," *New York Times,* 2010, http://www.nytimes.com/2010/12/09/garden/09Bees.html?scp=1&sq=legalizing%20beekeeping&st=cse (accessed February 2, 2011).

33. Joshua Brunstein, "Beekeepers Keep the Lid On," *New York Times,* 2009, http://www.nytimes.com/2009/06/21/nyregion/21ritual. html?scp=4&sq=legalizing%20beekeeping&st=cse (accessed February 3, 2011).

34. Ibid.

35. Colin Jerolmack, "Primary Groups and Cosmopolitan Ties: The Rooftop Pigeon Flyers of New York City," *Ethnography* 10, no. 4 (2009): 435–457.

36. Ralph Garner, "Generating Buzz and Honey," *Wall Street Journal,* 2010, http://online.wsj.com/article/SB10001424052748703860104575508042073485632.html?KEYWORDS=generating+buzz (accessed February 4, 2011).

37. Jen Wrye, "Beyond Pets: Exploring Relational Perspectives of Petness," *Canadian Journal of Sociology,* 34, no.4 (2009): 1041.

38. Lecture at the Brooklyn Brainary on July 9, 2011.

39. Jamie Gross, "That Buzzing Could Sweeten Tomorrow's Tea," *New York Times,* 2010, http://www.nytimes.com/2010/03/26/dining/26sfdine.html (accessed February 4, 2011).

40. Dick Hebdige, *Subculture: The Meaning of Style* (New York: Routledge, 1978).

41. Christian Lorentzen, "Why the Hipster Must Die," *Time Out New York,* http://www.timeout.com/newyork/things-to-do/why-the-hipster-must-die (accessed May 16, 2012).

42. For example, while the Williamsburg condo complex called "The Edge" was under construction it used signs advertising potential renters could have "indie bands and stone countertops" and "the hippest dress code and the coolest zip code." Fieldnotes, Mary Kosut.

43. http://www.helpthehoneybees.com/#helping.

44. George Ritzer, *Enchanting a Disenchanted World: Revolutionizing the Means of Consumption* (Thousand Oaks, CA: Sage, 2004).

45. Paul Robbins, ed., *Green Consumerism: An A-to-Z Guide* (Thousand Oaks, Calif.: Sage, 2010), 5.

46. "Save the Whale, Save the Planet," *BBC News*, 2000, http://news.bbc.co.uk/2/hi/uk_news/787425.stm (accessed December 8, 2011).

47. Wrye, "Beyond Pets," 1042.

48. From the World Wildlife Fund website, http://www.worldwildlife.org/species/flagship-species.html (accessed September 9, 2011).

CHAPTER 4

1. Bee sting statistics are available at Buzzaboutbees.net, http://www.buzzaboutbees.net/bee-sting-facts.html (accessed June 7, 2012).

2. See Werner Herzog's documentary *Grizzly Man* as an example of crossing human/animal boundaries and interspecies intimacy.

3. Retrieved from the website of the Center for Biological Diversity, http://www.biologicaldiversity.org/species/invertebrates/index.html (accessed June 18, 2012).

4. Josie Glausiusz, "Your Body Is a Planet," *Discover*, June 2007, http://discovermagazine.com/2007/jun/your-body-is-a-planet (accessed August 4, 2012).

5. Jane Bennett, *Vibrant Matter: A Political Ecology of Things* (Durham: Duke University Press, 2010), 112.

6. See, for example, Hugh Raffles, *Insectopedia* (New York: Pantheon, 2010); Jake Kosek, "Ecologies of Empire: On the New Uses of the Honeybee," *Cultural Anthropology* 25, no. 4 (2010): 650–678; Charles Zerner, "Stealth Nature: Biomimesis and the Weaponization of Life," in *The Name of Humanity: The Government of Threat and Care*, ed. Ilana Feldman and Miriam Tictin (Durham: Duke University Press, 2011).

7. S. Eben Kirksey, "Interspecies Love in an Age of Excess: Being and Becoming with a Common Ant, *Ectatomma ruidum* (Roger)," in *The Politics of Species: Reshaping Our Relationships with Other Animals*, ed. Raymond Corbey and A. Lanjouw (Cambridge: Cambridge University Press, 2013).

8. Ibid.

9. Bennett, *Vibrant Matter*, 23.

10. Rudolf Steiner, *Bees: Lectures by Rudolf Steiner*, translated by Thomas Braatz (Barrington, Mass.: Anthroposophic Press, 1998), 17.

11. Mary Kosut, "Extreme Bodies/Extreme Culture," in *The Body Reader*, ed. Lisa Jean Moore and Mary Kosut (New York: NYU Press, 2010), 184–200.

12. Eva Crane, *The Rock Art of Honey Hunters* (Cardiff, Wales: International Bee Research Association, 2001).

13. Stephen Buchmann, *Letters from the Hive: An Intimate History of Bees, Honey, and Humankind* (New York: Bantam, 2005), 15.

14. Steiner, *Bees*, 3.

15. Ibid.

16. Ibid.

17. Anders Blok, "Actor-Networking Ceta-Sociality, or, What Is Sociological about Contemporary Whales?" *Distinktion*, no. 15 (2007): 70.

18. Kennan Ferguson, "I © My Dog," *Political Theory* 32, no. 3 (2004): 373–395 (quote on 379).

19. Jen Wrye, "Beyond Pets: Exploring Relational Perspectives of Petness," *Canadian Journal of Sociology* 34, no. 4 (2009): 1035.

20. Patricia Clough, "(De) Coding the Subject-in-Affect," *Subjectivity* 23 (2008): 141.

21. Patricia A. Adler and Peter Adler, *The Tender Cut: Inside the Hidden World of Self-Injury* (New York: NYU Press, 2011).

CHAPTER 5

1. Lisa Jean Moore, *Sperm Counts: Overcome by Man's Most Precious Fluid* (New York: NYU Press, 2007).

2. For an example of a similar treatment of wildlife narration, see Cynthia Chris, *Watching Wildlife* (Minneapolis: University of Minnesota Press, 2006).

3. William Longgood, *The Queen Must Die: And Other Affairs of Bees and Men* (New York: Norton, 1988).

4. Ibid., 109.

5. The full *Oxford English Dictionary* record lists 1609 as the first mention of Queen[e] Bee. See Charles Butler, 1609, "Feminine Monarchie, or A treatise concerning bees," First Edition, Volume 1, i. sig. A3: Also, a definition for a queen bee can be found through the *OED*, where insects were referred to as queens in 1609 and as humans in 1780 or 1790 as "the chief or dominant woman in an organization or social group." http://www.oed.com/viewdictionaryentry/Entry/267967 (accessed August 15, 2012).

6. Susan Brackney, *Plan Bee: Everything You Ever Wanted to Know about the Hardest Working Insect on the Planet* (New York: Perigee, 2009), 15.

7. Donna Haraway, *The Companion Species Manifesto* (Chicago: Prickly Paradigm Press, 2003), 9.

8. See http://www.merriam-webster.com/dictionary/drone (accessed October 18, 2012).

9. Moore, *Sperm Counts*.

10. G. J. Barker-Benfield, "The Spermatic Economy: A Nineteenth Century View of Sexuality," *Feminist Studies* 1, no. 10 (1972): 45–74.

11. Longgood, *The Queen Must Die*, 116.

12. United States Congress, Majority Staff of the Joint Economic Committee, December 2010, http://www.jec.senate.gov/public/?a=Files.Serve&File_id=9118a9ef-0771-4777-9c1f-8232fe70a45c (accessed October 18, 2012).

13. See Daniel Swift, "Drone Knowns and Drone Unknowns," *Harper's Magazine*, October 2011, http://harpers.org/archive/2011/10/hbc-90008288 (accessed June 12, 2012); and Daniel Swift, "Conjectural Damage: A History of Bombing," *Harper's Magazine*, November 2011, http://harpers.org/archive/2011/11/0083690 (accessed June 12, 2012).

14. Jake Kosek, "Ecologies of Empire: On the New Uses of the Honeybee," *Cultural Anthropology* 25, no. 4 (2010): 650–678.

15. John Mello, "Computers Controlling Military Drones Reportedly Infected with Virus," *PC World*, 2011, http://www.pcworld.com/article/241507/computers_controlling_military_drones_reportedly_infected_with_virus.html (accessed October 18, 2012).

16. Arlie Hochschild, *The Second Shift: Working Parents and the Revolution at Home* (New York: Penguin, 1989).

17. Anna Tsing, "Empowering Nature, or: Some Gleanings in Bee Culture," in *Naturalizing Power: Essays in Feminist Cultural Analysis*, ed. Sylvia Yanagisako and Carol Delaney (New York: Routledge, 1995), 113–144.

18. Irene Cheng, "The Beavers and the Bees," *Cabinet* 23 (2006), http://www.cabinetmagazine.org/issues/23/cheng.php (accessed October 18, 2012).

19. Charlotte Sleigh, *Six Legs Better: A Cultural History of Mymecology* (Baltimore: Johns Hopkins University Press, 2007), 210.

20. For example, see Jack A. Wilson, "Ontological Butchery: Organism Concepts and Biological Generalizations," *Philosophy of Science* 67 (2002), http://joelvelasco.net/teaching/systematics/wilson%2000%20-%20ontological%20butchery.pdf (accessed June 14, 2012).

CHAPTER 6

1. Patrick Hedlund, "Massive Bee Swarm Shuts Down Little Italy Corner," *DNAinfo.com*, 2011, http://www.dnainfo.com/20110531/lower-east-side-east-village/massive-bee-swarm-shuts-down-little-italy-sidewalk (accessed October 12, 2012).

2. Garth Johnston, "Beelieve It or Not, NYC Has Its Own Bee Rescue Team," *The Gothamist*, 2011, http://gothamist.com/2011/05/28/beelieve_it_or_not_new_york_city_ha.php (accessed June 11, 2012).

3. Ibid.

4. For examples of flash mobs and crowd experiments, visit www.improveverywhere.com.

5. Eugene Thacker, "Networks, Swarms, Multitudes," *CTheory*.net, 2004, http://www.ctheory.net/articles.aspx?id=422 (accessed October 1, 2012).

6. Andre Stipanovic, "Bees and Ants: Perceptions of Imperialism in Vergil's Aeneid and

Georgics," in *Insect Poetics*, ed. Eric C. Brown (Minneapolis: University of Minnesota Press, 2006).

7. Jake Kosek, "Ecologies of Empire: On the New Uses of the Honeybee," *Cultural Anthropology* 25, no. 4 (2010): 650–678.

8. Tammy Horn, *Bees in America: How the Honey Bee Shaped a Nation* (Lexington: University Press of Kentucky, 2005).

9. The National Honey Board's website is http://www.honey.com/.

10. Kim Flottum, "2009 Has the Worst Honey Crop on Record," *The Daily Green*, 2009, http://www.thedailygreen.com/environmental-news/blogs/bees/honey-bee-keeping-47102806 (accessed June 14, 2012).

11. Michel Foucault, *"Society must be defended": Lectures at the Collège de France, 1975–1976*, translated by David Macey (New York: Picador, 2003).

12. Benjamin P. Oldroyd and Siriwat Wongsiri, *Asian Honey Bees: Biology, Conservation, and Human Interactions* (Cambridge: Harvard University Press, 2006).

13. See Andrews Schneider, "Asian Honey, Banned in Europe, Is Flooding the U.S. Grocery Shelves," *Food Safety News*, 2011, http://www.foodsafetynews.com/2011/08/honey-laundering/ (accessed June 12, 2012); and Bruce Boynton, "National Honey Board: Honey Is Made from Nectar, Not Pollen.," *Food Safety News*, 2012, http://www.foodsafetynews.com/2012/04/national-honey-board-honey-is-made-from-nectar-not-pollen/ (accessed October 20, 2012).

14. Kent Conrad, "Conrad Calls on FDA to Block Tainted Honey," press release, 2007, http://www.conrad.senate.gov/pressroom/record.cfm?id=276745 (accessed June 12, 2012).

15. Ibid.

16. Andrew Schneider, "Tests Show Most Store Honey Isn't Honey," *Food Safety News*, 2011, http://www.foodsafetynews.com/2011/11/tests-show-most-store-honey-isnt-honey/ (accessed June 12, 2012).

17. Read more at "Chinese Honey Now Reported among Import Dangers," *WorldNetWeekly*, 2007, http://www.wnd.com/?pageId=42270#ixzz1YoTykc86 (accessed June 11, 2012).

18. Raymond Williams, *Key Words: A Vocabulary of Culture and Society* (New York: Oxford University Press, 1972).

19. Vandana Shiva, *Stolen Harvest: The Hijacking of the Global Food Supply* (Boston: South End Press, 2000); Sarah Franklin and Celia Roberts, *Born and Made: An Ethnography of Preimplantation Genetic Diagnosis* (Princeton, N.J.: Princeton University Press, 2006).

20. Thomas Rinderer, "Russian Honey Bee Earning Its Stripes," United States Department of Agriculture, 2004, http://www.ars.usda.gov/is/AR/archive/oct01/bee1001.htm (accessed June 12, 2012).

21. Michael Zimmerman, *Contesting Earth's Future: Radical Ecology and Postmodernity* (Berkeley: University of California Press, 1997).

22. S. Hubbell and P. M. Breeden, "Maybe the 'Killer' Bee Should Be Called Bravo Instead," *Smithsonian* 22, no. 6 (1991).

23. Mark Winston, *Killer Bees: The Africanized Honey Bee in the Americas* (Cambridge: Harvard University Press, 1992), 5.

24. As Winston points out, "In hindsight, the importation of Africanized bees should not have taken place—or at least, stock should have been properly selected, bred, and tested prior to importation. The bee's reputation as a good honey producer has proven unfounded, due to a combination of high swarming and absconding rates and the unwillingness of most beekeepers to perform even minimal management in the face of serious stinging problems. By any economic, agricultural, public or political measure this importation was not desirable. In a biological sense, though, the bee has been highly successful, spreading at high rates and forming a dense feral population which may be having considerable impact on resident bees. Ironically, the characteristics that have proven deleterious for beekeeping ideally pre-adapted the African honeybee for a feral existence in South America" (Winston, *Killer Bees*, 15).

25. Hubbell and Breeden, "Maybe the 'Killer' Bee Should Be Called Bravo Instead," 116.

26. Brian Massumi, "The Future Birth of the Affective Fact," Conference Proceedings: *Genealogies of Biopolitics* (2005), http://browse.reticular.info/text/collected/massumi.pdf (accessed October 12, 2012).

27. Ibid., 8.

28. Achilles Mbembe, "Necropolitics," *Public Culture* 15, no. 1 (2003): 11–40.

29. Anna Tsing, "Empowering Nature, or: Some Gleanings in Bee Culture," in *Naturalizing Power: Essays in Feminist Cultural Analysis*, ed. Sylvia Yanagisako and Carol Delaney (New York: Routledge), 135.

30. Peggy Pascoe, *What Comes Naturally: Miscegenation Law and the Making of Race in America* (Oxford: Oxford University Press, 2009), 1.

31. Ibid., 307.

32. Winston, *Killer Bees*, 71.

33. Ibid., 81.

34. Angela Davis, *Women, Race, and Class* (New York: Vintage, 1983).

35. Thomas Dyer, *Theodore Roosevelt and the Idea of Race* (New Orleans: Louisiana State University Press, 1992).

36. Hugh Raffles, *Insectopedia* (New York: Pantheon, 2010), 144.

37. See Cori Hayden, *When Nature Goes Public: The Making and Unmaking of Bioprospecting in Mexico* (Princeton, N.J.: Princeton University Press, 2003); and Sarah Franklin, *Dolly Mixtures. The Remaking of Genealogy* (Durham: Duke University Press, 2007).

38. We take this term "production animals" from Hans Harbers, "Animal Farm Love Stories: About Care and Economy," in *Care in Practice: On Tinkering in Clinics, Homes and Farms*, ed. Annemarie Mol, Ingunn Moser, and Jeannette Pols (London: Transaction, 2010), 141–170.

39. Jacqueline Stevens, *Reproducing the State* (Princeton, N.J.: Princeton University Press, 1999), 269. Stevens demonstrates how the state establishes and re-establishes "orderly and disorderly forms of being of its population—through the reproduction of kinship rules on birth certificates, marriage licenses, passports, and so forth" (xv).

40. Ibid., 173.

41. See, for example, Michael O'Malley, *The Wisdom of Bees: What the Hive Can Teach Business about Leadership* (New York: Portfolio Hardcover, 2010).

42. Nicholas De Genova, "The Production of Culprits: From Deportability to Detainability in the Aftermath of 'Homeland Security,'" *Citizenship Studies* 11, no. 5 (2007): 421–448 (quote on 425).

43. Arnold Arluke, "A Sociology of Sociological Animal Studies," *Society and Animals* 10, no. 4 (2002): 369–374 (quote on 370).

CHAPTER 7

1. Marjorie Spiegel, *The Dreaded Comparison: Human and Animal Slavery* (Ann Arbor: University of Michigan Books, 1996); and Charles Patterson, *Eternal Treblinka: Our Treatment of Animals and the Holocaust* (New York: Lantern, 2002).

2. Everett Oretel, "History of Beekeeping in the United States," *Bee Source*, 1980, http://www.beesource.com/resources/usda/history-of-beekeeping-in-the-united-states/ (accessed June 12, 2012).

3. Jacob Lienbenluft, "Rent-a-Hive: How Much Does It Cost to Borrow a Colony of Honeybees?" *Slate*, 2008, http://www.slate.com/articles/news_and_politics/explainer/2008/06/rentahive.html (accessed August 4, 2012).

4. Timothy Pachirat, *Every Twelve Seconds: Industrialized Slaughter and the Politics of Sight* (New Haven: Yale University Press, 2011).

5. Jennifer Gregory Miller, "Catholic Activity: Celebrating the Feast of St. Ambrose," *Catholic Culture*, 2003, http://www.catholicculture.org/culture/liturgicalyear/activities/view.cfm?id=965 (accessed September 15, 2012).

6. In his lectures from 1923, Rudolf Steiner examines the workings of beehives as they relate to human health. See Rudolf Steiner, *Bees: Lectures by Rudolf Steiner*, translated by Thomas Braatz (Barrington, Mass.: Anthroposophic Press, 1998), 4.

7. Dan Charles, "Funny Honey? Brining Trust to a Sector Full of Suspicion," *The Salt*, December 2011, http://www.npr.org/blogs/thesalt/2011/12/13/142903171/funny-honey-bringing-trust-to-a-sweet-sector-fraught-with-suspicion (accessed January 15, 2012).

8. Thorstein Veblen, *The Theory of the Leisure Class: An Economic Study of Institutions* (New York: Macmillan, 1899).

9. Staff, "Bees Recognize Human Faces Using Feature Configuration," *Science Daily*, 2010, http://www.sciencedaily.com/releases/2010/01/100129092010.htm (accessed October 12, 2012).

10. Timothy Morton. *The Ecological Thought* (Cambridge: Harvard University Press, 2010).

11. Alex Williams, "Buying into the Green Movement," *New York Times*, 2007, http://www.nytimes.com/2007/07/01/fashion/01green.html?pagewanted=print&_r=0 (accessed June 1, 2012).

12. Zygmunt Bauman, *Work, Consumerism and the New Poor* (Philadelphia: Open University, 1998).

13. Dennis Soron, "Sustainability, Self-Identity and the Sociology of Consumption," *Sustainable Development* 18 (2010): 172–181 (quote on 177).

14. Craig J. Thompson and Gokcen Coskuner-Balli, "Enchanting Ethical Consumerism: The Case of Community Supported Agriculture," *Journal of Consumer Culture* 7, no. 3 (2007): 275–303.

15. Clive Hamilton, "Consumerism, Self-Creation and Prospects for a New Ecological Consciousness," *Journal of Cleaner Production* 18, no. 6 (2010): 571–575 (quote on 573).

16. David Kiley, "Best Global Brands," *Business Week* (August 7, 2006): 59.

17. Louise Story, "Can Burt's Bees Turn Clorox Green?" *New York Times*, 2008, http://www.nytimes.com/2008/01/06/business/06bees.html?pagewanted=all (accessed May 1, 2012).

18. See the Burt's Bees website and in particular their "story," http://www.burtsbees.com/c/story/mission-vision/what-we-re-not.html (accessed June 12, 2012).

19. Judith Falon, *Negotiating the Holistic Turn: The Domestication of Alternative Medicine* (Albany: SUNY Press, 2005).

20. Christopher Kim, "Apitherapy: Bee Venom Therapy," in *Potentiating Health and the Crisis of the Immune System: Integrative Approaches to the Prevention and Treatment of Modern Diseases*, ed. S. Fulder, A. Mizrahi, and N. Sheinman (New York: Springer, 1997), 243–270.

21. See the American Apitherapy Society website and in particular their mission statement at http://www.apitherapy.org/about-aas/mission-statement/ (accessed June 1, 2012).

22. Caitlin Berrigan, "The Life Cycle of a Common Weed: Viral Imaginings in Plant-Human Encounters," *WSQ* 40, no. 1/2 (2012): 97–116.

23. V. Robson, S. Dodd, and S. Thomas. "Standardized Antibacterial Honey (Medihoney) with Standard Therapy in Wound Care: Randomized Clinical Trial," *Journal of Advanced Nursing* 65, no. 3 (2009): 565–575.

24. A. Zumla and A. Lulat, "Honey—a Remedy Rediscovered," *Journal of the Royal Society of Medicine* 82 (1989): 384–385; and F. I. Seymour and K. S. West, "Honey—Its Role in Medicine," *Med Times* 79 (1951): 104–107.

25. Staff, "Manuka Honey Research to Grow Industry," NZNewsUK, 2011, http://www.nznewsuk.co.uk/news/?id=18578&story=Manuka-honey-research-to-grow-industry (accessed October 11, 2012)

26. Staff, "NZ Short on Beekeepers," Scoop: Independent News, 2011, http://www.scoop.co.nz/stories/BU1106/S00077/nz-short-on-beekeepers.htm (accessed October 12, 2012).

27. Aubrey Fine, *Handbook on Animal-Assisted Therapy: Theoretical Foundations and Guidelines* (Amsterdam: Elsevier, 2001).

28. See the American Veterinary Medical Association's Guidelines for Animal-Assisted Activity, Animal-Assisted Therapy, and Resident Animal Programs at https://www.avma.org/KB/Policies/Pages/Guidelines-for-Animal-Assisted-Activity-Animal-Assisted-Therapy-and-Resident-Animal-Programs.aspx (accessed October 20, 2012).

29. Tzachi Zamir, "The Moral Basis of Animal-Assisted Therapy," *Society and Animals* 14, no. 2 (2006): 179–199 (quote on 179).

30. For some examples of bee-venom therapy how-to videos, see http://www.youtube.com/watch?v=-N7JqkJ2nc0 and http://www.youtube.com/watch?v=FDT_Hwp8TSQ (accessed March 12, 2012).

31. Tom Regan, *Defending Animal Rights* (Urbana: University of Illinois Press, 2001).

32. Jeffrey Lockwood, *Six-Legged Soldiers: Using Insects as Weapons of War* (Oxford: Oxford University Press, 2010), 191.

33. Hugh Raffles, *Insectopedia* (New York: Pantheon, 2010), 191.

34. Hilda Ransome, *The Sacred Bee in Ancient Times and Folklore* (Mineola, N.Y.: Dover, 1937).

35. G. C. Rains, J. K. Tomberlin, and D. Kulasiri, "Using Insect Sniffing Devices for Detection," *Trends in Biotechnology* 26, no. 6 (2008): 288–294.

36. See, for example, Monica J. Casper and Lisa Jean Moore, "Dirty Work and Deadly Agents: (Dis)Embodied Politics of a Weapons Treaty," *WSQ* 39, no. 1–2 (2011): 95–119.

37. DARPA's continued involvement is uncertain. As reported in a 2006 news report, " 'Despite the positive test results, DARPA said it does not see a future for bomb-detecting bees in the military. Bees are not reliable enough for military tactical use at this point,' the agency said. . . . 'We see no clear pathway to make them reliable enough to make it worth risking the lives of our service men and women.' " In a follow-up interview, a DARPA spokeswoman, Jan Walker, said, "We're done in this research area. We don't plan any further investment." See Associated Press, "The Bombs and the Bees: Trained Bees Sniff out Explosives," *Tampa Bay Times*, December 9, 2006, http://www.sptimes.com/2006/12/09/Worldandnation/The_bombs_and_the_bee.shtml (accessed June 11, 2012).

38. See these videos for explanations of bees' military deployment: http://www.youtube.com/watch?v=_T7d0bze4kM and http://www.culanth.org/?q=node/370 (accessed March 12, 2012).

39. Lois Ember, "Bees on Patrol: Studies Find Bees Are Potential Chemical, Biological

Agent Detectors," *Chemical and Engineering News*, May 20, 2002, http://pubs.acs.org/cen/critter/bees.html (accessed April 27, 2010).

40. As Jerry Bromenshenk and his team have pointed out, "Bees do not fly at night, during heavy rain or wind, or when temperatures drop to near or below freezing. As such, the use of bees is seasonal in temperate climates. Bees are active year-round in tropical regions. These limitations are not unique. Dogs do not perform well in wind, rain or frozen ground, and dogs and handlers usually do not work in the dark. Unlike a dog, bees do not need to bond to their handlers, and they work whenever weather conditions permit. At a weight of one-tenth of a gram, bees are not going to trip wires or set any mines off. Their wide foraging range offers the possibly of greatly speeding up survey times, while also increasing

handler safety. Bee colonies can be established along the perimeter of the minefield, not in the minefield. With no leash to hold, the beekeepers can stay well clear of the dangerous area.

Bees have several advantages in addition to their keen sense of smell and wide area coverage:

Bees can be conditioned and put into use in one to two days.
Local bees and beekeepers are used.
Overall costs are far lower than for dog teams.
Bees are essential to re-vitalizing agriculture in war-torn countries."

See Jerry Bromenshenk, Colin Henderson, Robert Seccomb, Steven Rice, et al., "Can Honey Bees Assist in Area Reduction and Landmine Detection?" *Research, Development and Technology in Mine Action* 7, no. 3 (2003), http://maic.jmu.edu/journal/7.3/focus/bromenshenk/bromenshenk.htm (accessed October 12, 2012).

41. One new innovation is the manufacturing of mechanical bee eyes: cyborg trans-species technologies or bee eye optics simulations (BEOS). For more information, see W. Sturzl, N. Boeddeker, L. Dittmar, and M. Egelhaaf, "Mimicking Honeybee Eyes with 280-Degree Field of View Catadioptric Imaging System," *Bioinspiration and Biomimetics* 5, no. 3 (2010): 1–13, or http://iopscience.iop.org/1748–3190/5/3/036002/pdf/1748–3190_5_3_036002.pdf (accessed October 12, 2012).

42. Thomas D. Seeley, *Honeybee Democracy* (Princeton, N.J.: Princeton University Press, 2010).

43. Raffles, *Insectopedia*, 172.

44. The U.S. Department of Agriculture's Animal Welfare Act can be found here: http://awic.nal.usda.gov/nal_display/index.php?info_center=3&tax_level=3&tax_subject=182&topic_id=1118&level3_id=6735&level4_id=0&level5_id=0&placement_default=0 (accessed June 1, 2012).

45. Colin McGinn, "Animal Minds, Animal Morality," *Social Research* 62, no. 3 (1996): 731-747 (quote is on 737).

46. Cary Wolfe, *What Is Posthumanism*? (Minneapolis: University of Minnesota Press, 2009).

CHAPTER 8

1. Donna Haraway, *When Species Meet* (Minnesota: University of Minnesota Press, 2007), 46.

2. Steve Goodman, *Sonic Warfare: Sound, Affect and the Ecology of Fear* (Cambridge: MIT Press, 2010), 189.

3.There are many who have carefully thought through these issues philosophically and theoretically, see Isabelle Stengers and Robert Bononno, *Cosmopolitics* (Minneapolis: University of Minnesota Press, 2011); Karen Barad, *Meeting the Universe Halfway* (Durham: Duke University Press, 2007); Haraway, *When Species Meet*; Carol J. Adams, *Neither Man Nor Beast: Feminism and the Defense of Animals* (New York: Continuum, 1995); and Lori Gruen, *Ethics and Animals: An Introduction* (New York: Cambridge University Press, 2011), among many others.

INDEX

ABOUT THE AUTHORS

Lisa Jean Moore is a feminist medical sociologist at Purchase College, State University of New York. Her previous work explores human bodies and body parts and fluids in sociocultural contexts. She lives in Crown Heights, Brooklyn, with her husband, Paisley Currah, and their three daughters, two dogs, three fish, and two Xenopus (aquatic frogs).

Mary Kosut is a cultural sociologist and Associate Professor of Media, Society, and the Arts and Gender Studies at Purchase College, State University of New York. Her current research explores the sociality of art scenes in New York City. She lives in Bushwick, Brooklyn.